Marie Claire Cantone • Christoph Hoeschen

Editors

Radiation Physics
for Nuclear Medicine

 Springer

Editors

Prof. Dr. Marie Claire Cantone
Università degli Studi di Milano
Dept. of Physics
Via Celoria
20133 Milano
Italy
marie.cantone@unimi.it

Prof. Dr. Christoph Hoeschen
Helmholtz Zentrum München
Deutsches Forschungszentrum
für Gesundheit und Umwelt (GmbH)
Ingolstädter Landstr. 1
85764 Neuherberg
Germany
Christoph.hoeschen@helmholtz-muenchen.de

ISBN 978-3-642-11326-0   e-ISBN 978-3-642-11327-7
DOI 10.1007/978-3-642-11327-7
Springer Heidelberg Dordrecht London New York

Library of Congress Control Number: 2011921696

Cover design: eStudio Calamar, Figueres/Berlin

Printed on acid-free paper

Springer is part of Springer Science+Business Media (www.springer.com)

In Memory of Niky Molho (1938–1993)

The training course held in Milan on November 2008 on Radiation Physics in Nuclear Medicine, on which basis this book is conceived, was dedicated to memory of Niky Molho, on the occasion of the fifteenth anniversary of his death. Niky Molho was Professor of Medical Physics at the Università degli Studi di Milano (State University of Milan), Faculty of Medicine, and leader of the Research Unit on Medical Physics at the Department of Physics of the same University.

It was February 21, 1993 when a sudden death took Niky away from his dedicated and passionate teaching and research activities at the age of 55.

After receiving the University degree in Physics in 1962, he started to direct his research interests in the field of experimental nuclear physics, with particular attention for studies concerning reaction mechanisms, models of interpretation, gamma spectroscopy and instrumentation. From 1970 when he started to teach physics to the medical students of the Faculty of Medicine in Milan, he started also to focus his research activity toward the medical field, by making full use of his accumulated experiences and knowledge in nuclear physics. He started the development of a new methodology, based on nuclear activation by mean of proton beams for the quantitative analysis of trace elements in biological samples. Such methodology was used with great success for the determination of oligo-elements in human blood. On these premises he started a very close cooperation with the Institute of Biophysical Radiation Research of the GSF in Frankfurt am Main, Germany. This collaboration has been kept alive by his former students and collaborators until these days, although the German counterpart has now changed its name (Helmholtz Zentrum München, German Research Center for Environmental Health, Department of Radiation Sciences) and location (Munich).

Also as a result of this cooperation, he then oriented his efforts to the investigation of the metabolism of oligo-elements by using stable isotopes as tracers, opening a new field of research of fundamental importance for potential applications in various medical fields, including radiation protection. The result of the biokinetic studies that were initiated by Niky more than 25 years ago are actually currently employed for the validation and the revision of the biokinetic models for selected radionuclides presented by the International Commission on Radiological Protection, ICRP.

Beyond his scientific contributions, Niky has to be remembered as a passionate teacher at the Faculty of Medicine in Milan, where he became full professor in 1987. Over the years he also supervised a significant number of diploma theses in physics and master theses in medical physics at the University of Milan with dedication and contagious enthusiasm.

In parallel with his research and teaching activities, he dedicated a great deal of time and efforts to the promotion of the Medical Physics at all levels, since he was really convinced of the importance of this interdisciplinary branch of Physics. He has been actively involved in the most important initiatives in this field in Italy and in Europe. An example of this activity is the organization of the first course of EFOMP Advanced School of Medical Physics in Como, Italy, May 1992, dealing with 'Metabolism Studies using Magnetic Resonance Spectroscopy and Positron Emission Tomography', with participants and students from all over Europe.

As colleagues and students of those days, now active in various positions in Italy and in Europe, we all remember very well how Niky was constantly available and attentive to our concerns, questions and needs. In particular, he was able to establish an open mind dialogue with young students in physics, even on difficult and controversial matters, as only, an enthusiastic and passionate teacher can do. Many of us will always look respectfully at his rigor him as a precious "second father".

Marie Claire Cantone

Contents

Part I
Introduction

Chapter 1
The Role of Radiation Physics in Nuclear Medicine

Marie Claire Cantone and Christoph Hoeschen

Nuclear medicine is a fast-growing and dynamic specialty; in recent years with the widespread use of positron emission tomography (PET), PET combined with computed tomography (CT), single-photon emission computed tomography (SPECT)-CT, and new applications in radionuclide therapy, this discipline asks for continuous updating of nuclear medicine professionals on various aspects of the changing trends. Nuclear medicine has grown from the initial in vitro tests to advanced methods, allowing the imaging of organ function and thus becoming an important diagnostic tool. In fact, until now no other method than scintigraphy has been available to evaluate myocardial blood flow in stress and rest conditions or to evidence bone metastases.

Just by looking at the historical timeline of nuclear medicine, it can be seen that a number of important key improvements are related to the approaches for proper radionuclide production technologies; to the radiation detection methodologies and improved detector performances with respect to specific requirements in medicine; and to the use of more specific biokinetic models. Moreover, the main reason behind the constant increase in reliability of nuclear medicine is the continuous research and improvements in the field of image reconstruction. Nuclear medicine is a great area where knowledge and research not only in physics but also in chemistry and pharmacology meet knowledge and research in medicine resulting in a unique and expanding field of knowledge thanks to these synergies.

It is worth remembering that with the development of reactors and accelerators by physicists in the 1940s, artificial radionuclides were produced and prepared. Tc-99m, which is still one of the most used radionuclides in nuclear medicine, was discovered by the Italian physicist Emilio Segre in 1938. I-131 became the prototype for tissue-specific radiotherapy, and the treatment of hyperthyroidism is considered

M.C. Cantone (✉)
Università degli Studi di Milano, Italy
e-mail: marie.cantone@unimi.it

C. Hoeschen
Helmholtz-Zentrum München, Germany
e-mail: Christoph.hoeschen@helmholtzmuenchen.de

M.C. Cantone and C. Hoeschen (eds.), *Radiation Physics for Nuclear Medicine*,
DOI 10.1007/978-3-642-11327-7_1, © Springer-Verlag Berlin Heidelberg 2011

to be the beginning of nuclear medicine. As more knowledge was gained about basic biochemical processes, the use of radioactive tracers in diagnostic medicine became dramatically attractive. Scintillation counters such as NaI(Tl) and CsI(Tl) crystals were developed by the late 1940s to allow highly efficient photon detection, and in some cases solid-state detectors such as Si(Li) and Ge(Li) were developed by physicists to allow smaller detectors with better counting statistics. Originally, a single probe was set up over the patient's body or organ site using cylindrical collimation. Between 1952 and 1958, the scintillation camera, known as the anger camera, was gradually developed; through the imaging of the gamma rays emitted by the radionuclides used, it contributes to the medical diagnoses of tumours and other pathologies. The introduction of this easy-to-use imaging device, which is still the main instrument used all over the world, has brought a considerable focus on diagnostics by imaging.

During the 1960s, the growth of nuclear medicine was phenomenal. Its novelties and advances were strictly correlated to the improvement in technology and instrumentation based on the fundamental processes of radiation physics.

One of the most remarkable steps in the evolution of nuclear medicine has been the improvement in tomographic imaging. By starting from the mathematical analyses by the Austrian mathematician Johann Radon and physicist Allan McLeod Cormack and from the applied research of radiologist David Kuhl, the commercial practical application for the prototype of a CT scanner was patented in the early 1970s by Godfrey Hounsfield. The Nobel Prize in Medicine and Physiology was shared in 1979 by physicists Cormack and Hounsfield.

By the 1970s, most of the body organs could be visualized by nuclear medicine procedures, including brain tumour localization, liver and spleen scanning, and studies of the gastrointestinal tract. Moreover, the accelerator production of ^{18}F and of other positron emitted radionuclides enabled the growth of PET imaging. By 1971, nuclear medicine was recognised as a medical specialty by the American Medical Association, and toward the end of the 1980s, Rb-82 was approved as the first radiopharmaceutical for myocardial perfusion by PET. By the 1990s, PET was becoming one of the most important diagnostic tools for detailed images. The recent history of PET is characterised by continuous improvements in terms of resolution and sensitivity of the devices.

In the present environment, in which the latest results of research are becoming immediately available, this book presents the most recent advancements in nuclear medicine. It starts with the fundamental processes related to the interaction of radiation with matter; it describes the research and development in radiopharmaceuticals and in detection areas, and it introduces the new frontiers in nuclear medicine.

The book is developed in five sections. The introduction includes the contribution by Fridtjof Nüsslin, which focuses on the molecular imaging pathway with regard to biomedical physics.

In Part II "Fundamental Processes of Radiation Physics," the principles and concepts that are the building blocks of this discipline are discussed. The interaction between radiation and matter is addressed with special attention towards critical or complex aspects with regard to the interaction of charged particles and the Monte Carlo method, as used in simulation codes for radiation transport.

In Part III "Radiation Sources and Radiopharmaceutical Productions," the physics of isotope production, including radioisotope isolation and purification, manufacturing, and quality control of radiopharmaceuticals, is presented. Moreover, research and development of new radiopharmaceuticals and related special aspects of clinical development are considered.

In Part IV "Radiation Detectors for Medical Applications," the basic principles of detection for ionizing radiation used in medical imaging and the statistical and electronic treatment of the measurements are introduced. The characteristics of the most common semiconductors and scintillators currently used in medical imaging systems are described, and an overview of the main research areas in detector development and novel techniques currently investigated is presented.

In Part V "New Frontiers in Nuclear Medicine," the concept of the PET magnifier probe as well as the latest advancements in the fields of algorithms for image reconstruction, biokinetic models, and voxel phantoms is introduced.

Chapter 2
The Molecular Imaging Pathway to Biomedical Physics

Fridtjof Nüsslin

2.1 Introduction

According to Policy Statement 1 issued by the International Organization for Medical Physics (IOMP) [1], "Medical physics is a branch of physics that is concerned with the application of physics concepts and techniques to the diagnosis and treatment of human diseases." Although including the many areas in which medical physicists are not dealing with ionizing radiation (e.g. physiology, neurology, audiology), the well-known domains of medical physicists are radiotherapy, radiological diagnostics, and nuclear medicine. Traditional medical physics as a branch of applied physics has been driven mainly by the physical roots, that is, development of instrumentation and measurement techniques, specifically in the radiological fields concerning quality assurance, optimization of radiation applications, modelling, and simulation. With the rapid progress of biology and its ever-increasing role in all medical sciences, medical physics also is challenged to incorporate more and more biological knowledge, particularly what is designated with the buzz terms *molecular biology* or *molecular medicine*. In fact, nuclear medicine has been per se a "molecular" science since its beginnings, with radioisotopes used to bind to specific biomolecules either as biomarkers or as therapeutic weapons to attack malignant cancer cells, for instance. However, in radiotherapy and radiodiagnostics, the clinical impact of molecular biology is still limited; during the last decade, this has significantly increased. In particular, the potential use of molecular biology for the entire radiation treatment process can hardly be overestimated. Accordingly, also the medical physicist cannot ignore the current trend of integrating more biology knowledge in daily work, particularly when including molecular imaging data in the treatment optimization process. Actually, medical physics is expanding towards biomedical physics, as shown by the increasing number of departments and institutions with "biomedical physics" included in their name. It is molecular imaging that has an impact on radiotherapy and hence challenges the medical physicist to

F. Nüsslin
Klinikum rechts der Isar, Technische Universität München, Munich, Germany
e-mail: nuesslin@lrz.tu-muenchen.de

M.C. Cantone and C. Hoeschen (eds.), *Radiation Physics for Nuclear Medicine*,   7
DOI 10.1007/978-3-642-11327-7_2, © Springer-Verlag Berlin Heidelberg 2011

appreciate the new face of traditional medical physics, more appropriately termed *biomedical physics*.

As an example of the closer entanglement of medical physics and biology, this chapter illustrates some obvious opportunities and promises of the molecular approach of radiotherapy and its impact on medical physics or, better, biomedical physics.

2.2   Fundamentals of Molecular Imaging

Molecules are the building blocks of all life. Any failures within the molecular and cellular processes may lead to disorders of organs, organ systems, and normal tissue. Still a major life threat, cancer may originate from just a single aberrant molecule of a single cell. Medical interventions at the earliest stage therefore present a challenge detection of such molecular aberrations. Prevention, diagnosis, and therapy of cancer are hoped to be most effectively performed at the molecular level.

The term *molecular imaging* was coined by Weissleder and Mahmood [2]. Accordingly, molecular imaging aims to investigate the molecular signature of diseases through "in-vivo characterization and measurement of biologic processes at the cellular and molecular level." Molecular imaging addresses any molecular modification at the genomic, transcriptomic, proteomic, metabolomic levels, ranging from subcellular and cellular structures to the tissue and finally the organ.

Some biological processes accessible via imaging and relevant in cancer diagnosis and therapy are glycolysis, DNA synthesis, amino acid and lipid metabolism, apoptosis, tumour angiogenesis, and hypoxia.

Molecular imaging is based on the selection of an adequate technique combined with a suitable contrast agent. The most appealing imaging techniques are by far positron emission tomography (PET) and magnetic resonance (MR). Remarkable progress has been observed in ultrasound and optical imaging technologies, such as fluorescence absorption, reflectance, and luminescence. The characteristics of some imaging systems with particular relevance for clinical application are summarized in Table 2.1.

To increase efficiency, hybrid imagers have been developed, such as PET combined with computed tomography (CT), which combines high-resolution CT anatomical imaging and high-specificity and -sensitivity PET biological imaging in the same system. In the clinical environment, PET-CT imagers are invaluable in localization, treatment planning, and treatment monitoring for cancer. Further clinical hybrid scanners are single-photon emission computed tomography (SPECT) combined with CT and, most recently, MR-PET systems.

The visualization of aberrant molecules or molecules that are an integral part of a certain biological process needs essentially imaging methods with a high degree of specificity and sensitivity. Therefore, it is the target that defines the selection of the contrast agent. Thus, the specificity of molecular imaging is to a great deal dependent on the imaging agents, which in turn depend on the imaging system. Widely

Table 2.1 Characteristics of some molecular imaging systems with direct impact on clinical application

Feature	Ultrasound	CT	MRI	SPECT	PET
Image acquisition method	Reflexion at interfaces	X-ray attenuation	Electromagnetic excitation of nuclei; relaxation	Single-photon emission of radioiso-topes	β^+-Decay of radioiso-topes, annihilation radiation
Spatial resolution	1–3 mm	30 μm	<30 μm – 1 mm	0.3–1.5 mm	1–4 mm
Acquisition time	10–30 s	1 s per scan	1–10 min	10–60 min	10–80 min
Sensitivity	Good	Low	Low/medium	High	Very high
Specificity	Low	Low	Medium	High	Very high

CT computed tomography, *MRI* magnetic resonance imaging, *PET* positron emission tomography, *SPECT* single-photon emission computed tomography.

used molecular imaging agents are radiopharmaceuticals, paramagnetic materials, fluorescent/luminescent materials, and microbubbles. Reporter gene imaging, which provides data on the location, duration, and extent of gene expression, is rapidly progressing, particularly in biomedical research. More recent molecular imaging modalities are based on nanoparticle vehicles, smart contrast agents, and target-specific optical or radiolabelled agents.

2.3 Application of Biological Imaging in the Radiotherapy Process

There is clinical evidence that small cancers that have not yet metastasized can be cured by local treatment (i.e., surgery or radiotherapy or combined modalities). In this regard, the potential of molecular imaging is early detection of cancer. A successful cancer treatment requires elimination of all stem cells of a particular tumour whilst minimizing the radiation damage to the surrounding normal tissue. Hence, the strategy of radiation therapy is to modulate the delivered dose distribution correspondingly. Combining anatomical and biological imaging (multimodal imaging) is essential in all four phases of the radiotherapy treatment process: (1) tumour staging, (2) target volume definition, (3) treatment delivery and monitoring, and (4) treatment response. This sequence may be considered as a feedback loop in which tumour staging is the prerequisite for the choice of the most efficient therapy scheme, and the response to the selected treatment allows assessment of the prognosis of the disease. In clinical practice, contouring of target volumes benefits most from combining, for instance, CT and PET data sets. A second area in which only molecular imaging provides the essential information is biological tumour heterogeneity. Modern linear accelerators equipped with multileaf collimators are capable of applying intensity-modulated radiation therapy (IMRT) beams.

Ling [3] pioneered the concept of biological adapted therapy, that is, to modulate the dose distribution in the target corresponding to the profile of varying radiation sensitivity throughout the tumour. The essential steps of this concept, often called *dose painting* or *dose sculpting*, are (1) selecting appropriate biomarkers, (2) imaging the biological tumour subvolumes, (3) transforming the image data sets into a dose prescription, (4) optimizing the dose distribution and finally (5) delivering the modulated dose distribution. Hypoxia is well established, causing resistance in radiation therapy. In many malignancies, hypoxia has been identified as a major negative prognostic factor for tumour progression. Therefore, enhancing the dose locally at hypoxic tumour subvolumes is expected to lead to better treatment results. As surrogate markers for tumour hypoxia, PET imaging ^{18}F-Misonidazole (Miso), ^{18}F-labeled azomycin-arabinoside (AZA), and ^{60}Cu-labeled diacetyl-bis (N4-methylthiosemicarbazone) (ATSM) are used, and the first clinical studies have been just initiated. Dose prescription is difficult, but by applying the linear-quadratic model, an assessment can be made. The last step, transforming activity to dose distribution, is typically based on biological models. In general, there is rather poor information about the dynamics of the temporal variation of the tracer distribution. According to our own measurements on patients with head and neck tumours, there are significant changes during the initial phase immediately after injection of the tracer that require at least about 5 min to achieve some balancing. Translation of the concept of biological-driven dose modulation into clinics is not a common standard yet.

Treatment monitoring and response assessment are some of the most promising features of molecular imaging. The benefits for clinical outcome are the adaption of the treatment to the response, to indicate earlier whether a tumour responds to treatment, to monitor receptor expression with specifically labelled receptor ligands, and to monitor tumour response by labelling cell metabolites. Traditionally, standard CT anatomic evaluation of a tumour treatment needs several months to identify morphological changes. In contrast, when measuring the ^{18}F-fluorodeoxyglucose uptake it is possible to detect tumour regression or recurrence in just a few hours.

2.4 Conclusion

Molecular imaging as the hallmark of individualized therapy becomes more and more important in radiotherapy, ranging from early tumour detection via biologically and physically optimized treatment planning, image-guided treatment delivery, and early response assessment of the treatment. Accordingly, this trend challenges the medical physicist to incorporate more biological knowledge into his or her clinical work. Traditional medical physics is slowly mutating towards biomedical physics. Current education and training schemes for medical physicists should be adapted properly to include more biology and imaging technology.

References

1. International Organization for Medical Physics, Policy Statement 1, (2010), http://www.iomp. org.
2. R. Weissleder and U. Mahmood, Molecular imaging, Radiology 2119 (2001), 316.
3. C. C. Ling, J. Humm, S. Larson, H. Amols, Z. Fuks, S. Leibel, and J. A. Koutcher. Towards multidimensional radiotherapy (MD-CRT): biological imaging and biological conformality. Int. J. Radiat. Oncol. Biol. Phys. 47 (2000), 551.

Part II
Fundamental Processes on Radiation Physics

Chapter 3
Mechanisms of the Interactions Between Radiation and Matter

Giuseppe Battistoni

3.1 Energy Loss of Charged Particles

In principle, electromagnetic interactions are exactly calculable by quantum electro-dynamics. However, atomic and molecular physics introduce complexities that often eliminate the attempts of achieving an exact analytical formulation and require the introduction of approximantions or numerical approaches.

The most important atomic processes (bremsstrahlung excluded) undergone by charged particles when traversing media are related to Coulomb scattering of both atomic electrons and nuclei. Even though the basic process is always the same (i.e., the Coulomb scattering), because of the very different masses of electrons and nuclei, the effect is very different: the cross section is proportional to Z^2, and when expressed as a function of the transferred four-momentum q, it is approximately independent of mass. This means that for a given projectile the energy transfer is inversely proportional to the target mass. Therefore, interactions with electrons are by far the dominant source of charged particle energy losses, while they give a con-tribution proportional to the atomic number Z (that is, like Z independent point unit charges) to the angular deflection. Collisions with atomic nuclei result in negli-gible energy losses, but the angular deflection is roughly that of a single particle of charge Z; thus, it is proportional to Z^2. The angular deflection is associated mostly with atomic nuclei except for the lightest elements, for which the two contributions become comparable. A detailed discussion of all these topics can be found in the references.

Let us discuss separately the case of e^+e^- and of heavier charged particles, start-ing from these. The formula for the energy loss (dE/dx or "stopping power," i.e., energy loss per unit path length) of heavy particles is customarily derived using the distinction between "distant" and "close" collisions; these terms refer to the magni-tude of the momentum transfer and hence to the collision impact parameter (even though in a quantum mechanical sense). In distant collisions, the particle interacts

G. Battistoni
INFN, Sezione di Milano, Milano, via Celoria 16, 20133 Italy
e-mail: giuseppe.battistoni@mi.infn.it

M.C. Cantone and C. Hoeschen (eds.), *Radiation Physics for Nuclear Medicine*,
DOI 10.1007/978-3-642-11327-7_3, © Springer-Verlag Berlin Heidelberg 2011

with the atom as a whole; for close collisions, the interaction can be considered to be with free electrons, and atomic properties are not involved. Usually, the two contributions are computed considering all energy transfer T_e to atomic electrons below a given threshold η:

$$T_e < \eta \rightarrow \text{distant collisions}$$
$$T_e > \eta \rightarrow \text{close collisions}$$

The dE/dx related to distant collisions is computed according to the Bethe–Bloch approach, making use of the Born approximation and assuming the velocity of the incident projectile to be much larger than the velocity of atomic electrons. A full discussion can be found in Fano's 1963 review and other fundamental references [1–3].

The cross section for producing an electron of energy T_e for an incident particle of kinetic energy $T_0 = (\gamma - 1)Mc^2$ depends on the spin and is given by

$$\left(\frac{d\sigma}{dT_e}\right)_0 = \frac{2\pi z^2 r_e^2}{\beta^2} \frac{m_e c^2}{T_e^2} \left[1 - \beta^2 \frac{T_e}{T_{max}}\right] \tag{3.1}$$

for spin 0 particles and by

$$\left(\frac{d\sigma}{dT_e}\right)_{\frac{1}{2}} = \frac{2\pi z^2 r_e^2}{\beta^2} \frac{m_e c^2}{T_e^2} \left[1 - \beta^2 \frac{T_e}{T_{max}} + \frac{1}{2}\left(\frac{T_e}{T_0 + Mc^2}\right)^2\right] \tag{3.2}$$

for spin 1/2 particles; r_e is the classical electron radius.

The maximum energy transfer T_{max} to the electron is dictated by kinematics and given by

$$T_{max} = \frac{2m_e c^2 \beta^2 \gamma^2}{1 + 2\gamma \frac{m_e}{M} + \left(\frac{m_e}{M}\right)^2} \tag{3.3}$$

It is important to note that, apart from the extreme tail where $T \approx T_{max}$, the dependence on the energy of the secondary electron is essentially given by the $1/T^2$ term. It is therefore clear from such formulas that low-energy transfers are much more likely than large ones.

Let us recall the relations that link the four-momentum transfer to the target $\widetilde{q}$ to the energy transferred to a stationary target (the electron in this case) and to the scattering angle θ_{cms} in the center-of-mass system:

$$T_e = \frac{\widetilde{q}^2}{2m_e c^2}$$

$$p_{cms} = p_{lab} c \frac{m_e c^2}{\sqrt{s}} = \sqrt{\frac{\left[s - (Mc^2 + m_e c^2)^2\right]\left[s - (Mc^2 - m_e c^2)^2\right]}{4s}} \tag{3.4}$$

$$s = m_e^2 c^4 + M^2 c^4 + 2\gamma M m_e c^4$$

$$\left|\vec{q}_{\,\text{cms}}\right| \equiv \left|\vec{q}\right| \equiv q = 2\, p_{\text{cms}} c \, \sin\frac{\theta_{\text{cms}}}{2}$$

$$\widetilde{q}^{\,2} = \vec{q}_{\,\text{cms}}^{\,2} \equiv \vec{q}^{\,2} = 2\, p_{\text{cms}}^{2}\, c^{2}\,(1 - \cos\theta_{\text{cms}})$$

$$q_{\text{max}}^{2} = 2 m_e c^2 T_{\text{max}} = 4\, p_{\text{cms}}^{2}\, c^{2} = 4\frac{m_e^2 M^2 c^4 \beta^2 \gamma^2}{m_e^2 + M^2 + 2\gamma M m_e} = 4\frac{m_e^2 M^2 c^8 \beta^2 \gamma^2}{s} \tag{3.5}$$

and hence

$$\begin{aligned}
\beta^2 &= \frac{s p_{\text{cms}}^2 c^2}{m_e^2 M^2 c^8 + s p_{\text{cms}}^2 c^2} \\[4pt]
&= \frac{\left[s - \left(Mc^2 + m_e c^2\right)^2\right]\left[s - \left(Mc^2 - m_e c^2\right)^2\right]}{4 m_e^2 M^2 c^8 + \left[s - \left(Mc^2 + m_e c^2\right)^2\right]\left[s - \left(Mc^2 - m_e c^2\right)^2\right]}
\end{aligned} \tag{3.6}$$

$$\begin{aligned}
\gamma^2 &= \frac{s p_{\text{cms}}^2 c^2 + m_e^2 M^2 c^8}{m_e^2 M^2 c^8} \\[4pt]
&= \frac{\left[s - \left(Mc^2 + m_e c^2\right)^2\right]\left[s - \left(Mc^2 - m_e c^2\right)^2\right] + 4 m_e^2 M^2 c^8}{4 m_e^2 M^2 c^8}.
\end{aligned} \tag{3.7}$$

Then, the cross section for transferring $\tilde{q}$ to an electron for an incident particle of mass Mc^2, spin 0, and center-of-mass energy $\sqrt{s}$ can be transformed into

$$\left(\frac{d\sigma}{dq^2}\right)_0 = \frac{4\pi z^2 r_e^2}{\beta^2}\frac{m_e^2 c^4}{q^4}\left[1 - \beta^2 \frac{q^2}{q_{\text{max}}^2}\right] = \frac{4\pi z^2 \alpha^2 \left(\hbar c\right)^2}{\beta^2}\frac{1}{q^4}\left[1 - \beta^2 \frac{q^2}{q_{\text{max}}^2}\right] \tag{3.8}$$

and into

$$\left(\frac{d\sigma}{dq^2}\right)_{\frac{1}{2}} = \frac{4\pi z^2 r_e^2}{\beta^2}\frac{m_e^2 c^4}{q^4}\left[1 - \beta^2 \frac{q^2}{q_{\text{max}}^2} + \frac{1}{2}\left(\frac{T_e}{T_0 + Mc^2}\right)^2\right] \tag{3.9}$$

for spin 1/2 particles.

To obtain the distant collision contribution, one should sum over all possible atomic excitations with proper cross sections. To understand the problem, the incident projectile field can be assumed to "illuminate" the atom for a time $t = b/v$, where b is the impact parameter, therefore including all frequencies up to $1/t$. The contribution coming from distant collisions for energy transfers T_e such that $T_e < \eta$, where η is such that

- $\eta > I_i$, for whichever atomic level i ($i = K, L, M, \dots$)
- η such that $\hbar/\eta > r_{\text{atom}} \rightarrow$ the incident particle can be considered as a point one (3.1)

can express the stopping power as

$$\left(\frac{dE}{dx}\right)_{T_e < \eta} = \frac{2\pi z^2 r_e^2}{\beta^2} m_e c^2 n_e \left[\ln \frac{2m_e \beta^2 c^2 \eta}{(1 - \beta^2) I^2} - \beta^2\right], \qquad (3.10)$$

where r_e is the classical electron radius, n_e is the number of electrons per cubic centimeter in the medium, $n_e = N_A Z \rho / A$, N_A = Avogadro's number, where use has been made of the Born approximation and of the assumption that the velocity of the incident projectile is much larger than the velocity of atomic electrons.

I is the *mean excitation energy*, and it is the main parameter characterizing the material properties. It is a suitable (logarithmic) average over all possible atomic levels:

$$\ln I = \sum_i f_i \ln I \qquad \sum_i f_i = 1, \qquad (3.11)$$

where f_i are the so-called oscillator strengths.

A possible practical and simple fit to the experimental values of I is given by

$$\begin{aligned}
\frac{I}{Z} &= 12 + \frac{7}{Z}\, eV, \quad I < 163\, eV \\
\frac{I}{Z} &= 9.76 + 58.8\, Z^{-1.19}\, eV, \quad I \geq 163\, eV.
\end{aligned} \qquad (3.12)$$

The contribution coming from close collisions can be (approximately) computed assuming collisions on free electrons and integrating the cross section from η to a maximum energy T_δ:

$$\begin{aligned}
\left(\frac{dE}{dx}\right)_{0,\eta < T_e < T_\delta} &= \int_\eta^{T_\delta} \left(\frac{d\sigma}{dT_e}\right)_0 n_e T_e dT_e \\
&= \frac{2\pi z^2 r_e^2}{\beta^2} m_e c^2 n_e \left[\ln \frac{T_\delta}{\eta} - \beta^2 \frac{T_\delta - \eta}{T_{max}}\right] \qquad (3.13)
\end{aligned}$$

for spin 0 particles and

$$\begin{aligned}
\left(\frac{dE}{dx}\right)_{\frac{1}{2},\eta < T_e < T_\delta} &= \int_\eta^{T_\delta} \left(\frac{d\sigma}{dT_e}\right)_{\frac{1}{2}} n_e T_e dT_e \\
&= \frac{2\pi z^2 r_e^2}{\beta^2} m_e c^2 n_e \left[\ln \frac{T_\delta}{\eta} - \beta^2 \frac{T_\delta - \eta}{T_{max}} + \frac{1}{4} \frac{T_\delta^2 - \eta^2}{(T_0 + Mc^2)^2}\right]
\end{aligned}$$

$$\qquad (3.14)$$

for spin 1/2 particles.

Summarizing, the formula for the average energy loss of particles much heavier than electrons and charge z can be expressed by

$$\left(\frac{dE}{dx}\right)_0 = \frac{2\pi n_e r_e^2 m_e c^2 z^2}{\beta^2} \left[\ln\left(\frac{2m_e c^2 \beta^2 T_{max}}{I^2 (1-\beta^2)}\right) - 2\beta^2 - 2\frac{C}{Z} - \delta\right] \quad (3.15)$$

for spin 0 particles and by

$$\left(\frac{dE}{dx}\right)_{\frac{1}{2}} = \frac{2\pi n_e r_e^2 m_e c^2 z^2}{\beta^2}$$

$$\times \left[\ln\left(\frac{2m_e c^2 \beta^2 T_{max}}{I^2 (1-\beta^2)}\right) - 2\beta^2 + \frac{1}{4}\frac{T_{max}^2}{(T_0 + Mc^2)^2} - 2\frac{C}{Z} - \delta\right]$$

$$(3.16)$$

for spin 1/2 particles.

The quantity δ is the so-called density correction, extensively discussed in the literature and connected with medium polarization, and C is the shell correction, which takes into account the effect of atomic bounds when the projectile velocity is no longer much larger than that of atomic electrons and hence the approximations under which the Bethe–Bloch formula has been derived break down. This correction becomes important at low energies. Let us examine in more detail these two corrections in the following section.

3.1.1 The Density Effect

The density effect is the reduction of the energy losses due to the polarization of the medium [4]. It has been systematically evaluated, and the results can be approximated by a fit as a function of η and introducing the variable $X = \ln\eta / \ln 10$:

$$\begin{array}{lll} 0 < X < X_0 & \delta(\eta) = 0 \\ X_0 < X < X_1 & \delta(\eta) = \ln\eta^2 + C + a(X_1 - X_0)^m \\ X > X_1 & \delta(\eta) = \ln\eta^2 + C. \end{array} \quad (3.17)$$

The parameters C, X_0, X_1, a, and m depend on the material and on its physical state (density, etc.). It is important to stress that for large energies,

$$\delta(\eta) \rightarrow \ln\eta^2 \quad (3.18)$$

partially suppressing the relativistic rise of dE/dx. Therefore, the most significant consequence of the density effect is a reduction of the stopping power of high-energy projectiles. This has been studied in detail by Sternheimer [5,6].

3.1.2 Shell Corrections

The quantity C is the sum of the corrections for each electron shell of the atom to the Bethe–Bloch expression [3,7]:

$$C = C_K + C_L + C_M + \cdots \tag{3.19}$$

The variation of C_is with velocity and atomic number can be computed. Each term is large and negative at very low velocities, but as the velocity increases, the sign changes. Each C_i passes through a maximum and then goes down. Furthermore, each C_i should approach zero as η^{-2} for large η. C_is also vary rapidly with the mean ionization energy I. A practical fit for C, valid for $\eta > 0.13$ can be expressed by

$$
\begin{aligned}
C(I,\eta) = &(0.422377\,\eta^{-2} + 0.0304043\,\eta^{-4} - 0.00038106\,\eta^{-6}) \cdot 10^{-6} \cdot I^2 \\
&+ (3.858019\,\eta^{-2} - 0.166798\eta^{-4} + 0.00157955\,\eta^{-6}) \cdot 10^{-9} \cdot I^3
\end{aligned}
\tag{3.20}
$$

where I is given in electron volts.

3.1.3 General Properties of Energy Loss

For most practical purposes, dE/dx in a given material depends only on the particle velocity β and charge z; thus, particles of the same velocity and charge have roughly the same energy loss. Moreover, if one measures distances in units of $\rho\,dx$ (g/cm^2), the energy loss is weakly dependent on the material as it proceeds like Z/A plus the logarithmic dependence on I. The proportionality to z^2 is evident.

The energy loss, when plotted as a function of $\beta\gamma = pc/M$, has a broad minimum at $\beta\gamma \approx$ 3–3.5. This minimum is almost constant up to very high energies if the restricted energy loss (i.e., the energy loss due to energy transfers smaller than some suitable threshold) is considered. In practice, most relativistic particles have energy losses in active detectors close to the minimum and are called *minimum ionizing particles* (MIPs).

3.1.4 Restricted Energy Loss

We can introduce the concept of *restricted energy loss*, considering the average energy loss of particles much heavier than electrons and charge z, with energy transfers to atomic electrons restricted at T_δ. This is given by

$$
\left(\frac{dE}{dx}\right)_{0,T_\delta} = \frac{2\pi n_e r_e^2 m_e c^2 z^2}{\beta^2}\left[\ln\left(\frac{2m_e c^2 \beta^2 T_\delta}{I^2(1-\beta^2)}\right) - \beta^2\left(1 + \frac{T_\delta}{T_{max}}\right) - 2\frac{C}{Z} - \delta\right]
\tag{3.21}
$$

for spin 0 particles and by

$$\left(\frac{dE}{dx}\right)_{\frac{1}{2},T_\delta} = \frac{2\pi n_e r_e^2 m_e c^2 z^2}{\beta^2} \left[\ln\left(\frac{2m_e c^2 \beta^2 T_\delta}{I^2(1-\beta^2)}\right) - \beta^2\left(1+\frac{T_\delta}{T_{max}}\right)\right.$$
$$\left. + \frac{1}{4}\frac{T_\delta^2}{(T_0+Mc^2)^2} - 2\frac{C}{Z} - \delta\right] \qquad (3.22)$$

for spin 1/2 particles.

When the density effect comes into play, $\delta \to \ln\left[\beta^2/(1-\beta^2)\right]$ and, for a given T_δ, the energy loss becomes constant.

3.1.5   Energy Loss of e^+e^-

As far as dE/dx for e^+ and e^- is concerned, there is no major difference with respect to the case of "heavy" particles, but for the differences in the "energetic" collisions with atomic electrons. The formulas for elastic Coulomb scattering for producing an electron of energy $T_e = \varepsilon T_0$ on free electrons by e^- and e^+ (Möller [8] and Bhabha [9] scattering) are given respectively by

$$\left(\frac{d\sigma}{dT_e}\right)_{\text{Möller}} = \frac{2\pi r_e^2}{\beta^2}\frac{m_e c^2}{T_e^2}\left[1 + \left(\frac{T_e}{T_0-T_e}\right)^2 + \left(\frac{\gamma-1}{\gamma}\right)^2\left(\frac{T_e}{T_0}\right)^2\right.$$
$$\left. - \frac{2\gamma-1}{\gamma^2}\left(\frac{T_e}{T_0-T_e}\right)\right] \qquad (3.23)$$

$$\left(\frac{d\sigma}{dT_e}\right)_{\text{Bhabha}} = \frac{2\pi r_e^2}{\beta^2}\frac{m_e c^2}{T_e^2}\left[1 - \frac{\gamma^2-1}{\gamma^2}\frac{T_e}{T_0} + \frac{1}{2}\left(\frac{\gamma-1}{\gamma}\right)^2\left(\frac{T_e}{T_0}\right)^2\right.$$
$$- \frac{\gamma-1}{\gamma+1}\frac{T_e}{T_0}\left\{\frac{\gamma+2}{\gamma} - 2\frac{\gamma^2-1}{\gamma^2}\frac{T_e}{T_0} + \left(\frac{\gamma-1}{\gamma}\right)^2\left(\frac{T_e}{T_0}\right)^2\right\}$$
$$+ \left(\frac{\gamma-1}{\gamma+1}\right)^2\left(\frac{T_e}{T_0}\right)^2$$
$$\left.\left\{\frac{1}{2} + \frac{1}{\gamma} + \frac{3}{2\gamma^2} - \left(\frac{\gamma-1}{\gamma}\right)^2\frac{T_e}{T_0}\left(1 - \frac{T_e}{T_0}\right)\right\}\right]. \qquad (3.24)$$

After integration on the proper kinematical limits (T_0 for positrons and $T_0/2$ for electrons), the resulting dE/dx is given by ($z^2 = 1$)

$$\left(\frac{dE}{dx}\right)_{e^-} = \frac{2\pi n_e r_e^2 m_e c^2}{\beta^2}\left[\ln\frac{T_0^2(\gamma^2-1)}{2I^2} + (1-\beta^2)^2\right.$$
$$\left. - \frac{2\gamma-1}{\gamma^2}\ln 2 + \frac{1}{8}\left(\frac{\gamma-1}{\gamma}\right)^2 - \delta\right] \qquad (3.25)$$

for electrons and

$$\left(\frac{dE}{dx}\right)_{e+} = \frac{2\pi n_e r_e^2 m_e c^2}{\beta^2} \left[\ln \frac{2T_0^2 (\gamma^2 - 1)}{I^2} \right.$$
$$\left. - \frac{\beta^2}{12} \left\{ 23 + \frac{14}{\gamma + 1} + \frac{10}{(\gamma + 1)^2} + \frac{4}{(\gamma + 1)^3} \right\} - \delta \right] \qquad (3.26)$$

for positrons.

3.1.6 The Case of Mixture and Compounds: The Bragg Additivity Rule

The expressions in the preceding section are all defined for a given element specified by the values of Z, A, and ρ. When a material is a mixture (compound) of several elements, in a first approximation the average energy loss can be computed according to the *Bragg additivity approximation* as

$$\left(\frac{dE}{d\rho x}\right)_{\text{tot}} = \frac{\sum_i \rho \left(\frac{dE}{d\rho x}\right)_i}{\rho_{\text{tot}}} \qquad (3.27)$$

In the case of compounds, this approximation is justified by the fact that aggregation (molecular binding) effects are usually negligible when the wavelength of the radiation is smaller than typical interatomic distances. Then, a molecule can be modeled as a set of individual, noninteracting atoms with uncorrelated positions. As a consequence of the Bragg additivity approximation, a "global" mean ionization energy $<I>$ can be defined according to

$$\ln \langle I \rangle = \frac{\frac{1}{\rho_{\text{tot}}} \sum_i \frac{Z_i}{A_i} \rho_i \ln I_i}{\langle \frac{Z}{A} \rangle}; \quad \left\langle \frac{Z}{A} \right\rangle = \frac{1}{\rho_{\text{tot}}} \sum_i \frac{Z_i}{A_i} \rho_i \qquad (3.28)$$

In a similar way, shell corrections can be approximately averaged as

$$\ln \left\langle \frac{C}{Z} \right\rangle = \frac{\frac{1}{\rho_{\text{tot}}} \sum_i \frac{Z_i}{A_i} \rho_i \frac{C_i}{Z_i}}{\langle \frac{Z}{A} \rangle} \qquad (3.29)$$

It has to be stressed again that both relations hold in an approximate manner only since chemical bounds, material state (solid, liquid, gas), general valence electron properties (metals, semiconductors, etc.), all affect both the mean ionization energy and the shell corrections with respect to the constituent elements. The determination of the actual I values to be used in compounds is a critical issue for precise calculations. A direct measurement is far from trivial. Of relevance for medical physics is the value for liquid water: International Commission on Radiation Units and Measurements (ICRU) report 49 [7] recommends the value $I = 75\pm3$ eV, but there exist now suggestions that higher values (in the range 80–85 eV) should be considered (see for instance [10]).

3.1.7 Other Corrections

For very low values of β, the Bethe–Bloch formula necessarily breaks down: the $1/\beta^2$ behavior cannot be extrapolated for $\beta \to 0$ easily. Actually, a number of complicated effects come into play when the velocity of the incident particle becomes comparable to that of atomic electrons. The most important effect is that the particle can pick up an atomic electron for part of the time; consequently the energy loss reaches a maximum and drops sharply. At present, the projectile charge distributions that cover a more or less noticeable range of ions, targets, and velocities are not available. Therefore, one can deal with various empirical and semiempirical fitting expressions for the average or, in other words, effective charge z_{eff}. The effective charge is to replace the bare projectile charge in all the relevant expressions. For protons and other single charged particles, the effective charge is assumed to be equal to the bare charge down to about 100 keV/A. For αs, a special fit by Ziegler [11] independent of the target material is used at all particle energies. For all the other ions, more elaborate fitting expressions that include a dependence on the target material are used.

In addition to this argument, at low energies the Born approximation is no more sufficient, and corrections terms proportional to z^3 must be considered (commonly known as Barkas corrections [12]). Notice that the addition of a term proportional introduces a (small) dependence on the sign of the charge (e.g., it gives rise to a small difference between protons and antiprotons). All these corrections must be considered for precise calculations at low velocity, in particular for particles close to the end of range, where the energy deposition reaches the maximum, known as the *Bragg peak*. Precise and detailed calculation of energy deposition as a function of penetration depth around the region of the Bragg peak can be achieved only with numerical or Monte Carlo calculations, and the reference values have been tabulated by ICRU [3, 7]. Examples of precise *dE/dx* calculations are given in Fig. 3.1 (Stopping and Range of Ions in Matter code [13]) and Fig. 3.2 (FLUKA Monte Carlo code [14]), where a $1/\beta^2$ functions are also superimposed to appreciate the deviation from the pure Bethe–Bloch function (Eq. 3.10). The examples of measured average stopping power for the case of heavy ions (argon and uranium) at different energies in different materials [15] as compared to Monte Carlo calculations using the FLUKA code [14] are shown in Fig. 3.3.

Figures 3.1 to 3.3 concern the average values of *dE/dx*, but the treatment of energy loss fluctuations is fundamental as well. This is addressed after a short discussion on the concept of range.

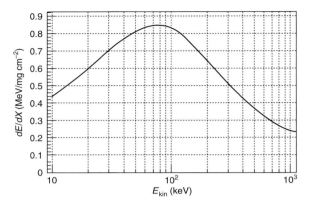

Fig. 3.1 Stopping power as a function of energy for protons in liquid water as calculated with the SRIM code [13]

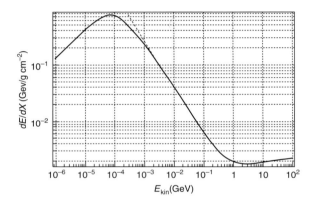

Fig. 3.2 Stopping power as a function of energy for protons in graphite as calculated with the FLUKA code [14]. The dotted line represents the back extrapolaton of the $1/\beta^2$ behavior of the Bethe–Bloch expression (Eq. 3.10)

3.1.8 Range of Particles

The total path traversed by a particle before coming at rest can be expressed by

$$R_{\text{true}} = \int\limits_{E_0}^{0} \left(\frac{dE}{dx} \right)^{-1} dE \qquad (3.30)$$

where R_{true} is a random variable characterized by its average value, variance, and so on.

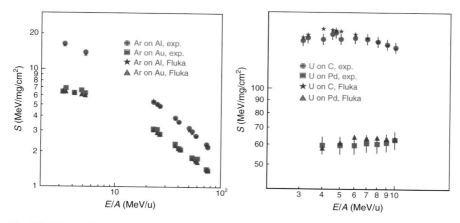

Fig. 3.3 Measured average stopping power as a function of energy/nucleon for argon ions in Au and Al [15] (*left*) and uranium ions in C and Pd (*right*), as compared to FLUKA [14] calculations

It is customary to introduce the so-called open "continuous slowing down approximation" (CSDA) range R_{csda}, or simply R, through

$$R_{csda} = \int_{E_0}^{0} \left(\frac{dE}{dx} \right)^{-1}_{mean} dE \qquad (3.31)$$

R_{csda} is the total amount of matter traversed by a particle of initial energy E_0 whenever the energy losses are equal to the average ones.

Since dE/dx is in a first approximation only a function of the particle velocity β and of its (squared) charge z, the following scaling properties hold:

$$R_b(M_b, z_b, p_b) = \left[\frac{M_b/M_a}{z_b^2/z_a^2} \right] R_a(M_a, z_a, p_a = p_b M_a/M_b). \qquad (3.32)$$

For example, the scaling relation predicts that the range of an α particle of momentum p_b is approximately equal to the range of a proton of momentum $p_b/4$, that is, of energy $T_p = T_a/4$ in the nonrelativistic regime and again $T_p \approx T_a/4$ in the relativistic one. Figure 3.4 shows the values of the CDSA range of electrons, protons, and α particles in liquid water as tabulated from information in [3, 7].

The range of particles is subject to important statistical fluctuations (straggling). Indeed, a precise prediction of the Bragg peak region requires a correct and complete treatment of the stochastic nature of the energy loss in the interactions with atoms. In practice, this can be accomplished by Monte Carlo calculations. As an introduction to this topic, we now summarize the few basic facts about statistical fluctuations in dE/dx.

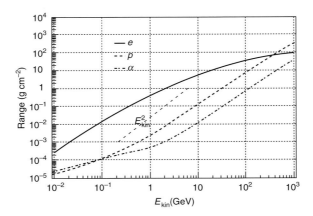

Fig. 3.4 Continuous slowing down approximation (CSDA) range as a function of energy for electrons, protons, and α particles in liquid water resulting from reference tabulations [3, 7]. A dotted line representing the $E_{kin}{}^2$ behavior has been superimposed for reference

3.1.9 δ-Ray Production

The energy lost by particles in atomic interactions results in excitation or ionization of the atoms. In case of ionizations, electrons are extracted; these are called δ rays. The productions of δ rays can be described by a cross section, which can be expressed in the following simplified form:

$$\frac{d\sigma}{dT} \approx \frac{2\pi z^2 r_e^2 m_e c^2}{\beta^2} \frac{1}{T^2} \tag{3.33}$$

Formula 3.33 is approximately valid for all particles.

The (average) number of δ rays with energy between T_{min} and T_{max} produced in a path length t such that the resulting energy loss is negligible compared with the initial particle energy is given by

$$\langle n_\delta \rangle = \sum_\delta t = n_e t \int_{T_{min}}^{T_{max}} \frac{d\sigma}{dT} dT = \frac{2\pi z^2 r_e^2 n_e m_e c^2}{\beta^2} t \left(\frac{1}{T_{min}} - \frac{1}{T_{max}} \right) \tag{3.34}$$

and the corresponding (average) energy loss by

$$\langle \Delta E_\delta \rangle = \langle T_\delta \rangle \langle n_\delta \rangle = n_e t \int_{T_{min}}^{T_{max}} T \frac{d\sigma}{dT} dT = \frac{2\pi z^2 r_e^2 n_e m_e c^2}{\beta^2} t \ln \frac{T_{max}}{T_{min}}. \tag{3.35}$$

The production of δ rays is the source of large fluctuations in the energy loss process.

The straggling in the δ energy loss distribution can be evaluated making use of a very general property of Poisson distributed variables. Given a Poisson distributed number of events n, each described by a variable x, with given $<n>$, $<x>$ and $<x^2>$, the following relations hold for the statistical variable $y = \sum_{i=0}^{n} x_i$:

$$\langle y \rangle = \langle n \rangle \langle x \rangle ; \quad \sigma_y^2 \equiv \langle y^2 \rangle - \langle y \rangle^2 = \langle n \rangle \langle x^2 \rangle. \tag{3.36}$$

Therefore,

$$\sigma_{\Delta E_\delta}^2 = \langle T_\delta^2 \rangle \langle n_\delta \rangle = n_e t \int_{T_{min}}^{T_{max}} T^2 \frac{d\sigma}{dT} dT = \frac{2\pi z^2 r_e^2 n_e m_e c^2}{\beta^2} t (T_{max} - T_{min}). \tag{3.37}$$

3.1.10 The Landau Fluctuations

It is useful to introduce the energy ξ such that the probability of emitting one δ ray with energy $T \geq \xi$ along t is one (it is customary to assume $T_{max} \gg \xi$ in the definition):

$$\xi = \frac{2\pi z^2 r_e^2 n_e m_e c^2}{\beta^2} t \tag{3.38}$$

Recalling that the maximum energy transfer to an atomic electron is given by Eq. 3.3, a parameter that is indicative of the skewness of the ΔE distribution can be built as

$$\kappa = \frac{\xi}{T_{max}} = \frac{\pi z^2 r_e^2 n_e}{\beta^4 \gamma^2} t \left[1 + 2\gamma \frac{m_e}{M} + \left(\frac{m_e}{M} \right)^2 \right]. \tag{3.39}$$

The parameter κ is governing the shape of the energy loss distribution. For $\kappa \gg 1$, the distribution is roughly Gaussian, while for $\kappa \ll 1$ (or equivalently $T_{max} \rightarrow \infty$) the distribution approaches the Landau one, which is defined by an integral of complex variable. For numerical purposes, a convenient form of the integral is given by

$$f(\lambda) = \frac{1}{\pi \xi} \int_0^\infty e^{-u \ln u - \lambda u} \sin \pi u \, du \tag{3.40}$$

where λ is defined in terms of the energy loss ΔE as

$$\lambda \equiv \frac{\Delta E - \Delta E_{mp}}{\xi}. \tag{3.41}$$

Here, ΔE_{mp} is the *most probable* energy loss value (i.e.s the peak value). This is different from the *average* energy loss $\langle E \rangle \approx \xi$. The two values are linked by the following relation:

$$\Delta E_{mp} = \xi \left[\ln \frac{\xi 2 m_e c^2 \beta^2 \gamma^2}{I^2} - \beta^2 + 1 - \gamma_E \right]. \qquad (3.42)$$

Here, $\gamma_E = 0.577\ldots$ is the Euler constant.

As stated, in the limit $\kappa \gg 1$, instead we get a Gaussian distribution, which can be expressed as

$$f(\Delta E) \approx \frac{1}{\xi \sqrt{\frac{2\pi}{\kappa} \left(1 - \frac{\beta^2}{2} \right)}} \exp \left[-\frac{(\Delta E - \langle \Delta E \rangle)^2 \kappa}{2\xi^2 \left(1 - \frac{\beta^2}{2} \right)} \right]. \qquad (3.43)$$

It is important to remember that δ ray production is not sufficient to give a full account of the *dE/dx* fluctuations. Also, atomic excitation (dominated by distant, soft collisions) is subject to fluctuations. This further straggling component has been added in the model of Blunk and Liesegang considering the convolution of the Landau function with a normal distribution with a variance related to soft collisions [16]. The inclusion of excitation straggling is customarily considered in refined Monte Carlo calculations. Figure 3.5 shows the comparison of experimental measurements [17] of energy loss distributions for 2-GeV/c positrons (left) and protons (right) traversing 100 μm of Si as compared with detailed simulations using the FLUKA Monte Carlo code [14], where a specific original model is adopted [18]. A pure Landau distribution would not fit the experimental data.

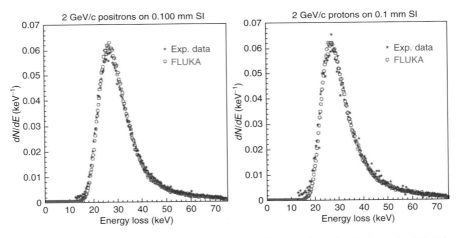

Fig. 3.5 Comparison of experimental measurements of energy loss distributions for 2-GeV/c positrons (*left*) and protons (*right*) traversing 100 μm of Si as compared with detailed simulations using the FLUKA Monte Carlo code [14, 18]

Another approximation in the derivation of the Landau distribution is the replacement of the actual value for the maximum energy loss in a single collision T_{max} with infinity. A more realistic description, using the actual value of T_{max}, has been obtained by Vavilov [19]. In the case of particles heavier than the electron, Vavilov's distribution provides a more realistic description of the energy loss fluctuations. Among the other features exhibited by this formulation, it must be mentioned that for $T_{max} \ll \xi$ Vavilov's distribution tends to the normal one, as expected from the central-limit theorem.

3.1.11 Multiple Coulomb Scattering

Another important aspect in precision calculations of charged particle interaction with matter is the deflection due to scattering processes with atoms. In practice, the goal is to calculate the angular distribution (and lateral displacement) of particles after traveling a certain path length in a given material (i.e., after many individual scatterings); therefore, a model for "multiple scattering" is necessary. This is a basic ingredient of all "condensed history" Monte Carlo algorithms.

As a starting point, we consider the Rutherford cross section for an incident particle of mass M and energy E against a nucleus of charge Z; it can be written as

$$\frac{d\sigma_{Ruth}}{d\Omega} = \frac{z^2 Z^2 r_e^2 m_e^2 c^4}{4\beta^4 E^2 \sin^4 \frac{\theta}{2}} = \left[\frac{z^2 Z^2 r_e^2}{4\beta^4 \gamma^2 \sin^4 \frac{\theta}{2}} \right] \left[\frac{m_e}{M} \right]^2. \tag{3.44}$$

In general, the Rutherford cross section requires further corrections, particularly when the condition $\pi \alpha z Z \ll 1$ is not verified (α is the fine structure constant, i.e., $e^2/\hbar c$). These corrections are important at relatively large angles:

$$\frac{d\sigma_{Coul}}{d\Omega} = \frac{d\sigma_{Ruth}}{d\Omega} \times G(z, Z, \beta). \tag{3.45}$$

The actual cross section for an atomic nucleus is given by

$$\frac{d\sigma_{Coul\,nuc}}{d\Omega} = \frac{d\sigma_{Coul}}{d\Omega} \times \frac{[F_{nuc}(q) - F_{at}(q)]^2}{Z^2} \times \left[F_{proj}(q)/z \right]^2 \tag{3.46}$$

where q is the momentum transfer, $F_{nuc}(q)$ is the target nucleus charge form factor, and $F_{at}(q)$ is the target nucleus elastic atomic form factor. $F_{proj}(q)$ is the projectile charge form factor (to be considered for ion projectiles). The target form factors are defined in such a way that

$$F_{at}(q) \xrightarrow{q \to 0} Z$$

$$F_{nuc}(q) \xrightarrow{q \to 0} z$$

$$F_{at}(q) \xrightarrow{q \gg \hbar / r_{at}} 0 \tag{3.47}$$

$$F_{nuc}(q) \xrightarrow{q \gg \hbar / r_{nuc}} 0.$$

In the case of the scattering with an atomic electron ($Z = 1$), Eq. 3.46 is replaced by

$$\frac{d\sigma_{\text{Coul nuc}}}{d\Omega} = \frac{d\sigma_{\text{Coul}}}{d\Omega} \times S_{at}(q) \times \left[F_{proj}(q)/z \right]^2 \tag{3.48}$$

where $S_{at}(q)$ is the inelastic atomic form factor:

$$S_{at}(q) \xrightarrow{q \to 0} 0$$

$$S_{at}(q) \xrightarrow{q \gg \hbar / r_{at}} Z. \tag{3.49}$$

For a generic value of q, an atomic model must be considered. This is the case in the Molière theory [20], where the cross section is eventually given by

$$\frac{d\sigma_{\text{Mol}}}{d\Omega} = \left[\frac{z^2 Z^2 r_e^2 m_e^2 c^4}{4\beta^4 E^2 \sin^4 \frac{\theta}{2}} \right] \left[\frac{(1 - \cos\theta)^2}{\left(1 - \cos\theta + \frac{1}{2}\chi_a^2\right)^2} \right] \tag{3.50}$$

where χ_a is a characteristic parameter, known as the *screening angle*, which in the Molière's computations is given by

$$\chi_a^2 = \frac{\chi_{cc}^2 e^{1 - 2\gamma_E}}{b_c E^2 \beta^2}; \quad \gamma_E = 0.577\ldots \tag{3.51}$$

Here, χ_{cc} and b_c are the only material-dependent quantities, given by

$$\chi_{cc} = \left[4\pi n z^2 Z \left(Z + \xi_e \right) r_e^2 m_e^2 c^4 \right]^{\frac{1}{2}} = \left[4\pi n z^2 Z \left(Z + \xi_e \right) \alpha^2 \left(\hbar c \right)^2 \right]^{\frac{1}{2}}$$

$$b_c = \frac{0.855^2 4\pi n z^2 Z \left(Z + \xi_e \right) \left(\hbar c \right)^2}{1.167 m_e^2 c^4 Z^{\frac{2}{3}} \left(1.13 + 3.76 \frac{\alpha^2 z^2 Z^2}{\beta^2} \right)} \tag{3.52}$$

where n is the number of atoms per unit volume; z and Z are the atomic numbers of projectile and target, respectively; ξ_e is a parameter taking into account the scattering on atomic electrons; and α is the fine structure constant.

In the small-angle approximation, the Molière cross section becomes

$$\frac{d\sigma_{small}}{d\Omega_{small}} = \left[\frac{4z^2 Z^2 r_e^2 m_e^2 c^4}{\beta^4 E^2 \theta^4}\right] \left[\frac{\theta^4}{\left(\theta^2 + \chi_a^2\right)^2}\right].$$

$$(d\Omega_{small} = 2\pi\theta d\theta) \tag{3.53}$$

Spin relativistic corrections should be taken into account; according to the second Born approximation formula (for spin 1/2 particles):

$$\frac{d\sigma}{d\Omega} = \frac{d\sigma_{Mol}}{d\Omega} = \left[1 - \beta^2 \sin^2 \frac{\theta}{2} - \pi z Z \alpha \beta \sin \frac{\theta}{2}\left(1 - \sin\frac{\theta}{2}\right)\right]. \tag{3.54}$$

On the basis of this processes, different multiple-scattering models have been developed. The fundamental approaches start from a simplified situation in which fast charged particles of a given kinetic energy E travel along a path length short enough so that their average energy loss is negligible compared to E. Here, we summarize the theories of Goudsmit and Saunderson [21] and Molière [20].

In the model of Goudsmit and Saunderson, the probability of scattering through an angle θ after traveling a total step length t (in g/cm^2) is expressed as an expansion of Legendre polynomials $P_l(\cos\theta)$:

$$P_{GS}(\theta, t) = \sum_{l=0}^{\infty}\left(l + \frac{1}{2}\right)e^{-tG_l}P_l(\cos\theta), \tag{3.55}$$

where G_l (transport coefficients) are obtained from the integration of the single scattering cross section $d\sigma/d\Omega$:

$$G_l = 2\pi\frac{N_A}{P_A}\int_0^\pi \frac{d\sigma(\theta')}{d\Omega}\left[1 - P_l(\cos\theta')\right]\sin\theta' d\theta' \tag{3.56}$$

N_A and P_A are the Avogadro's number and the atomic weight respectively. $G_0 = 0$ and G_l increases monotonically with l and tends to the value of the collision probability per unit path length when l goes to infinity. The Goudsmit–Saunderson expansion is exact, and the number of terms needed to get convergence of the series increases as the path length decreases. In practice, the problem is to calculate the G_l coefficients; in general, this requires a numerical approach. Because of the fast oscillations of Legendre polynomials, this is not a trivial calculation.

The approach of Goudsmit and Saunderson remains valid as long as the energy loss along the path t is neglected; therefore, it cannot be used for arbitrary values of the path length. The Molière approach to multiple scattering is instead based on approximate analytic expression, which results in a distribution that can be evaluated for arbitrary path lengths (larger than a few mean free paths). The Molière distribution can be easily sampled and therefore is used in several Monte Carlo codes because it is convenient in the simulations for which the path length t fluctuates randomly.

In the Molière model, the probability of scattering through an angle θ after traveling a total step length t is given by

$$F_{\text{Mol}}(\theta, t)\, d\Omega = 2\pi \chi d\chi \left[\frac{\sin\theta}{\theta}\right]^{\frac{1}{2}} \left[2e^{-\chi^2} + \frac{1}{B}f_1(\chi) + \frac{1}{B^2}f_2(\chi) + \cdots\right],$$

$$(3.57)$$

where χ is a scaled variable defined as

$$\chi = \frac{\theta}{\chi_c\sqrt{B}}; \quad \chi_c = \frac{\chi_{cc}\, t^{\frac{1}{2}}}{\beta^2 E}.$$

$$(3.58)$$

χ_{cc} has been introduced (Eq. 3.52), and B is the solution of the following transcendental equations:

$$B - \ln B = \ln \Omega_0; \quad \Omega_0 = \frac{b_c t}{\beta^2}.$$

$$(3.59)$$

Also, the parameter b_c was defined in Eq. 3.52.

The functions $f_n(\chi)$ are defined by the following integral expression containing the Bessel function J_0:

$$f_n(\chi) = \frac{1}{n!}\int_0^\infty u\, du\, J_0(\chi u)\, e^{-\frac{u^2}{4}}\left(\frac{u^2}{4}\ln\frac{u^2}{4}\right)^n.$$

$$(3.60)$$

An example of a Monte Carlo calculation using the Molière model is given in Fig. 3.6, which shows the calculated distribution of the scattering angle, projected onto a plane, for 100-MeV protons crossing a depth $t = 0.1\,\text{g/cm}^2$ of CaCO$_3$

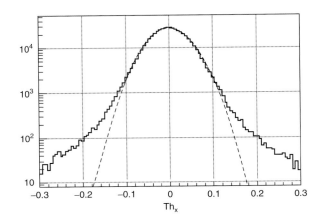

Fig. 3.6 Distribution of the scattering angle, projected on a plane, for 100-MeV protons crossing a depth $t = 0.1\,\text{g/cm}^2$ of CaCO$_3$ as calculated with the FLUKA code [14]. The dotted line represents, for reference, a simple Gaussian distribution

(FLUKA code). For reference, a pure Gaussian shape has been superimposed; notice in fact that the first term in the series of Eq. 3.57 is Gaussian. From the figure, we appreciate how multiple scattering exhibits tails that are much larger with respect to the Gaussian approximation.

A more elaborate theory of multiple scattering, accounting for the effects of energy loss, has been developed by Lewis [22], making use of the CDSA approach. However, for practical purposes, the Molière model remains convenient, provided that careful sampling of the step length is performed.

Correct modeling of multiple scattering is of great importance in calculations for medical physics. In particular, as far has hadron therapy is concerned, it must be noticed that the largest scattering angles and integrated displacements occur in the region of the Bragg peak, therefore potentially affecting the tissues surrounding the tumor target.

References

1. Fano, U., Penetration of protons, alpha particles and mesons, Ann. Rev. Nucl. Sci. 13, 1–66 (1963).
2. Inokuti, M., Inelastic collisions of fast charged particles with atoms and molecules – The Bethe theory revisited, Rev. Mod. Phys. 43, 297–347 (1971).
3. International Commission on Radiation Units and Measurements, Stopping Powers for Electrons and Positrons, ICRU report 37 (1984).
4. Fermi, E., The ionization loss of energy in gases and in condensed materials, Phys. Rev. 57, 485–493 (1940).
5. Sternheimer, R. M., The density effect for the ionization loss in various materials, Phys. Rev. 88, 851–859 (1952); Erratum: The density effect for the ionization loss in various materials, Phys. Rev. 89, 1309 (1953).
6. Sternheimer, R. M., Seltzer, S. M. and Berger, M. J., Density effect for the ionization loss of charge particles in various substances, Phys. Rev. B26, 6067–6076 (1982); Erratum, Phys. Rev. B27, 6971 (1983); At. Data Nucl. Data Tables 26, 261–271 (1984).
7. International Commission on Radiation Units and Measurements, Stopping Powers and Ranges for Protons and Alpha Particles, ICRU report 49 (1993).
8. Möller, C., Zur Theorie des Durchgangs schneller Elektronen durch Materie, Ann. Physik. 14, 531–585 (1932).
9. Bhabha, H. J., The scattering of positrons by electrons with exchange on Dirac's theory of electrons, Proc. Phys. Soc. A. 154, 195–196 (1965).
10. Emfietzoglou, D., Pathak, A., Papamichael, G., Kostarelos, K., Dhamodaran, S., Sathish, N., and Moskovitch, M., A study on the electronic stopping of protons in soft biological matter, Nucl. Instr. Meth. B242, 55–60 (2006); Paul, H., The ratio of stopping powers of water and air for dosimetry application in tumor therapy, Nucl. Instr. Meth. Phys. Res. B 256, 561–564 (2007).
11. Ziegler, J. F., Biersack, J. P., and Littmark, U., The Stopping and Range of Ions in Solids, Pergamon Press, Oxford (UK) (1985).
12. Jackson, J. D. and McCarthy, R. L., z^3 corrections to energy loss and range, Phys. Rev. B6, 4131–4141 (1972).
13. Ziegler, J. F., Biersack, J. P., and Ziegler, M. D., SRIM the Stopping and Range of Ions in Matter (2008): see http://www.srim.org.
14. Ferrari, A., Sala, P. R., Fasso, A., and Ranft, J., FLUKA: a multi-particle transport code, CERN 2005–10, INFN/TC_05/11, SLAC-R-773 (2005); Battistoni, G., Muraro, S., Sala, P. R.,

Cerutti, F., Ferrari, A., Roesler, S., Fasso', A., and Ranft, J., The FLUKA code: Description and benchmarking, Proceedings of the Hadronic Shower Simulation Workshop 2006, Fermilab 6–8 September 2006, M. Albrow, R. Raja eds., AIP Conference Proceeding 896, 31–49 (2007).

15. Bimbot, R., Compilation, measurements and tabulation of heavy ion stopping data, Nucl. Instr. Meth. B69, 1–9 (1992).

16. Blunk, O. and Leisegang, S., Zum Energieverlust schneller Elektronen in dunnen Schichten, Z. Pjysick 128, 500–505 (1950).

17. Bak, J. F., Bureknov, A., Petersen, J. B. B., Uggerhøj, E., Møller, S. P., and Siffert, P., Large departure from Landau distributions for high energy particles traversing thin Si and Ge targets, Nucl. Phys. B. 288, 681–716 (1987).

18. Fasso, A., Ferrari, A., Sala, P. R., and Ranft, J., New developments in FLUKA modelling of hadronic and EM interactions, Proceedings of the SARE-3, KEK-Tsukuba, May 7–9 1997, H. Hirayama ed., KEK report Proceedings 97–5, 32–40 (1997).

19. Vavilov, P. V., Ionization losses of high-energy heavy particles, Sov. Phys. JEPT 5, 749–751 (1957).

20. Molière, G., Theorie der Streuung schneller gelander Teilchen I: Einzelstreuung am abgeschirmten Coulomb-field, Z. Naturforsch. 2a, 133–145 (1947); Molière, G., Theorie der Streuung schneller gelander Teilchen II: Merfach- und Vielfachstreuung, Z. Naturforsch. 3a, 78–97 (1948).

21. Goudsmith, S., and Sanderson, J. L., Multiple scattering of electrons, Phys. Rev. 57, 24–29 (1940); Goudsmith, S. and Sanderson, J. L., Multiple scattering of electrons II, Phys. Rev. 58, 36–42 (1940).

22. Lewis, H. W., Multiple scattering in an infinite medium, Phys. Rev. 78, 526–529 (1950).

Chapter 4
Principles of Monte Carlo Calculations and Codes

Alberto Fassò, Alfredo Ferrari, and Paola R. Sala

4.1 Introduction

The Monte Carlo method was invented by John von Neumann, Stanislaw Ulam, and Nicholas Metropolis (who gave it its name) and independently by Enrico Fermi. Originally, it was not a simulation method but a mathematical approach aimed at solving a multidimensional integro-differential equation by means of a stochastic process. The equation itself did not necessarily refer to a stochastic process.

When the method is applied to a physical stochastic process, such as neutron diffusion, the model (in this case a random walk) could be identified with the process itself. Under those circumstances, the Monte Carlo method represents a simulation technique since every step of the model is aimed at mimicking an identical step in the physical process.

Particle transport and interaction represent a typical stochastic process and is therefore perfectly suitable for Monte Carlo simulation. Many applications, especially in high-energy physics and medicine, are based on simulations in which the history of each particle (trajectory, interactions) is reproduced in detail. However, in other types of application, typically energy deposition studies, or shielding design, the user is interested only in the expectation values of some quantities (fluence and dose) at some space point or region, which are calculated as solutions of a mathematical equation.

In many practical problems, it is more efficient to substitute the actual process with a modified one, resulting in the same average values, by sampling from suitably modified distributions. Such a *biased* process, if based on mathematically correct variance reduction techniques, converges to the same expectation values as the

A. Ferrari (✉)
European Laboratory for Particle Physics (CERN), 1211 Geneva 23, Switzerland
e-mail: alfredo.ferrari@cern.ch

A. Fassò
SLAC National Accelerator Laboratory, 2575 Sand Hill Road, Menlo Park, CA 94025, USA

P.R. Sala
INFN Sezione di Milano, Via Celoria 16, 20133 Milan, Italy

M.C. Cantone and C. Hoeschen (eds.), *Radiation Physics for Nuclear Medicine*,
DOI 10.1007/978-3-642-11327-7_4, © Springer-Verlag Berlin Heidelberg 2011

unbiased one, albeit at a faster pace. The drawback is that it cannot provide information about fluctuations and correlations of statistical distributions. In addition, the faster convergence in some regions of phase space is compensated by slower convergence elsewhere.

Thanks to the availability of faster and cheaper computers and of computer codes of improved quality, an increasing number of particle transport calculations in scientific and applied fields are carried out by Monte Carlo programs. The main advantage is mainly their capability to handle problems of practically any degree of complexity.

While before the 1990s the typical Monte Carlo approach was to simplify problems as much as possible, the modern Monte Carlo codes instead require fewer approximations (an exception are the condensed histories, described in a further section) and provide more accurate solutions. The possibility to cope with problems that before could not be solved has opened the way to a large number of new applications.

4.2 Phase Space

The concept of phase space is central to the understanding of the Monte Carlo method. Phase space is a concept of classical statistical mechanics. Each phase space dimension corresponds to a particle degree of freedom: Three dimensions correspond to the position in (real) space (x, y, z), and three other dimensions correspond to the momentum: p_x, p_y, p_z (or to energy and direction: E, θ, ϕ). More dimensions may correspond to other possible degrees of freedom: quantum numbers (e.g., spin), particle type, and so on. Each particle is represented by a point in phase space. Time can also be considered a coordinate, or it can be considered an independent variable: The variation of the other phase space coordinates as a function of time (the trajectory of a phase space point) constitutes a particle "history."

4.2.1 Phase Space Density

The basic quantity describing a population of particles is the phase space density $n(t, x, y, z, p_x, p_y, p_z)$, which is the number of particles in an infinitesimal volume of phase space. However, from the point of view of particle transport, the product $n\vec{v}$ of the space phase density and the particle velocity is more important because it represents the rate of the path length density and therefore relates to the particle interaction rate with matter. The quantity $\Psi = n\vec{v}$ is called *angular flux* and is the most general radiometric quantity.

Ψ can also be defined as the derivative of the fluence Φ with respect to all phase space coordinates: time, energy, and solid angle (direction vector):

$$\Psi = \frac{\partial \Phi}{\partial t \, \partial E \, \partial \vec{\Omega}} = \dot{\Phi}_{E\vec{\Omega}} \,. \tag{4.1}$$

The angular flux is a fully differential quantity, but most Monte Carlo solutions are integrals of Ψ over one or more (or all) phase space dimensions: coordinates, time, energy, angle. Fluence Φ, on the other hand, is the most integral radiometric quantity:

$$\Phi = \int_E \int_{\vec{\Omega}} \int_t \dot{\Phi}_{E\vec{\Omega}} dE \, d\vec{\Omega} dt. \tag{4.2}$$

Often in Monte Carlo calculations, the time dependence is not explicitly followed, and the fluence differential with respect to energy $\Phi(E) = \frac{d\Phi}{dE}$ is the quantity of interest.

4.2.2 The Boltzmann Equation

The Boltzmann equation is a balance equation in phase space: At any phase space point, the increment of particle density n in an infinitesimal phase space volume is equal to the sum of all "production terms" minus the sum of all "destruction terms." The same balance equation can be written in terms of angular flux $\Psi = nv$:

$$\frac{1}{v} \frac{\partial \Psi(x)}{\partial t} + \vec{\Omega} \cdot \nabla \Psi(x) + \Sigma_t \Psi(x) - S(x)$$
$$= \int_\Omega \int_E \Psi(x) \Sigma_s(x' \to x) dx' \tag{4.3}$$

where x represents all phase space coordinates: $\vec{r}, \vec{\Omega}, E, t$.

The various elements of the Boltzmann equation have the following physical meaning:

- The $\frac{1}{v} \frac{\partial \Psi(x)}{\partial t}$ term represents the time-dependent change of angular flux, for instance, due to particle decay.
- $\vec{\Omega} \cdot \nabla \Psi(x)$ is the change due to translational movement without change of energy and direction.
- $\Sigma_t \Psi(x)$, where Σ_t is the total macroscopic cross section (inverse of the mean free path), is a term representing absorption.
- $S(x)$ represents the particle sources.
- The double integral, where Σ_s is the macroscopic scattering cross section, refers to scattering: change in angular flux due to direction (and possibly energy) change without a change of particle position.

All Monte Carlo particle transport calculations are attempts to solve the Boltzmann equation in its integral form, that is, carrying out the integration over all possible particle histories.

To solve the Boltzmann equation, and hence for all Monte Carlo simulations, it is required to define one or more sources and one or more detectors.

In the most general case, a source consists of one or more particle types arbitrarily distributed in space, energy, angle, and time. In the simplest case, a source is simply a monoenergetic, monodirectional point source, that is, a "pencil beam."

A detector also is a region of phase space in which to look for the solution of the Boltzmann equation. Solutions can be of different type: at a number of (real or phase) space points, averaged over (real or phase) space regions, time-dependent or stationary, and so on. More generally, a detector is defined by distributions of Ψ in some of the phase space coordinates and integrated over others. It is interesting to notice the symmetry between sources and detectors, and indeed in some low-energy Monte Carlo codes they can be exchanged (adjoint mode).

The user must define a detector for each solution requested.

4.3 The Mathematical Basis of the Monte Carlo method

Most of the theoretical and mathematical foundations of the Monte Carlo method, as well as most basic textbooks on that technique [1–5], were historically centered on low-energy neutron and photon transport. However, several modern programs have extended the same mathematical concepts to the transport of charged particles and to interactions at higher energies, with some necessary additions mentioned in further sections (condensed histories, decay, transport in electric and magnetic fields).

4.3.1 Mean of a Distribution

Given a variable x distributed according to a function $f(x)$, the mean or average of another function of the same variable $A(x)$ over an interval $[a, b]$ is given by

$$\bar{A} = \frac{\int_a^b A(x) f(x) \mathrm{d}x}{\int_a^b f(x) \mathrm{d}x} \tag{4.4}$$

Or, introducing the normalized probability density function (pdf) $f'(x)$:

$$f'(x) = \frac{f(x)}{\int_a^b f(x) \mathrm{d}x} \tag{4.5}$$

$$\overline{A} = \int_a^b A(x) f'(x) \mathrm{d}x \tag{4.6}$$

A special case is that of $A(x) = x$:

$$\overline{x} = \int_a^b x f'(x) \mathrm{d}x \tag{4.7}$$

The concept of the mean of a distribution can be easily generalized to many dimensions. Given n variables $x, y, \ldots$, distributed according to the pdfs $f'(x), g'(y)$,

$h'(z), \ldots$, the mean or average of a function of those variables $A(x, y, z, \ldots)$ over an n-dimensional domain is given by

$$\overline{A} = \int_x \int_y \int_z \cdots \int A(x, y, z, \ldots) f'(x) g'(y) h'(z) \ldots dx dy dz \ldots \qquad (4.8)$$

An n-dimensional integral is often impossible to calculate with traditional methods, but we can sample N values, $A_i(x_i, y_i, z_i, \ldots)$, of A with probability $f'(x_i) g'(y_i) h'(z_i) \ldots$ and divide the sum of the sampled values by N:

$$S_N = \frac{\sum\limits_1^N A_i(x_i, y_i, z_i, \ldots)}{N}. \qquad (4.9)$$

Since each term of the sum is distributed like A, in this case the integration is also a simulation (analog Monte Carlo).

4.3.2   Central Limit Theorem

The central limit theorem [6] says that for large values of N, the distribution of the averages (normalized sums S_N) of N independent random variables identically distributed (according to *any* distribution with mean $\overline{A}$ and variance $\sigma^2 \neq \infty$) tends to a normal distribution with mean $\overline{A}$ and variance σ_A^2 / N:

$$\lim_{N \to \infty} S_N = \lim_{N \to \infty} \frac{\sum\limits_1^N A_i(x_i, y_i, z_i, \ldots)}{N} = \overline{A} \qquad (4.10)$$

$$\lim_{N \to \infty} P(S_N) = \frac{1}{\sqrt{\frac{2\pi}{N}} \sigma_A} e^{-\frac{(S_N - \overline{A})^2}{2\sigma_A^2 / N}} \qquad (4.11)$$

The central limit theorem is the mathematical foundation of the Monte Carlo method. Given any observable A, that can be expressed as the result of a convolution of random processes, the average value of A can be obtained by sampling many values of A according to the probability density of the random processes.

The Monte Carlo method is indeed an integration technique that allows to solve multidimensional integrals by sampling from suitable stochastic distributions [7].

As a consequence, the accuracy of the Monte Carlo estimator depends on the number of samples:

$$\sigma \propto \frac{1}{\sqrt{N}} \qquad (4.12)$$

4.3.3 Analog Monte Carlo

In an analog Monte Carlo calculation, not only the mean of the contributions converges to the mean of the actual distribution but also the variance and all moments of higher order; that is, fluctuations and correlations are preserved in principle like in the actual physical process. In this case (and in this case *only*), we have a real simulation.

4.4 Integration by Monte Carlo

4.4.1 Integration Efficiency

Traditional numerical integration methods (e.g., the Simpson rule) converge to the true value as $N^{-1/n}$, where N is the number of integration points or intervals and n is the number of dimensions. Integration by Monte Carlo converges as $N^{-1/2}$, independent of the number of dimensions. Therefore, depending on the problem dimensionality n,

- $n = 1$: Monte Carlo is not convenient.
- $n = 2$: Monte Carlo is about equivalent to traditional methods.
- $n > 2$: Monte Carlo converges faster (and increasingly so the greater the dimensions).

The dimensions of the integro-differential Boltzmann equation are the seven dimensions of phase space.

4.4.2 Random Sampling

The use of random-sampling techniques is the distinctive feature of Monte Carlo. The central problem of the Monte Carlo technique is as follows: Given a probability density function (pdf) of the x variable, $f(x)$, generate a sample of x's distributed according to $f(x)$ (x can be multidimensional).

Solving the integral Boltzmann transport equation by Monte Carlo consists of two essential parts: describing the geometry and materials of the problem and sampling randomly the outcome of physical events from probability distributions.

4.4.2.1 Random and Pseudorandom Numbers

The basis of all Monte Carlo integrations is random values of a variable distributed according to a pdf. In the real physical world, an experiment samples a large number of random outcomes of physical processes; these correspond, in a computer calculation, to pseudorandom numbers (PRNs) sampled from pdf distributions. The basic pdf is the uniform distribution $f(\xi) = 1$, with $0 \leq \xi < 1$.

PRNs are sequences that reproduce the uniform distribution and are constructed from mathematical algorithms (PRN generators). A PRN sequence looks random, but it is not; it can be successfully tested for statistical randomness although it is generated deterministically. A pseudorandom process is easier to produce than a true random one and has the advantage that it can be reproduced exactly.

PRN generators have a period, after which the sequence is repeated. The length of the sequence must be much longer than the number of random numbers required in any calculation. The period of the generator by Marsaglia and Tsang [8] is more than 10^{61}, and that of the "Mersenne twister" by Matsumoto and Nishimura [9] is longer than $10^{6,000}$.

4.4.2.2 Sampling from a Discrete Distribution

Consider a *discrete* random variable x that can assume values $x_1, x_2, \ldots, x_n, \ldots$ with respective probabilities $p_1, p_2, \ldots, p_n, \ldots$. Assume $\Sigma_i p_i = 1$ or normalize it.

Let us divide the interval $[0, 1)$ in n subintervals, with limits $y_0 = 0, y_1 = p_1, y_2 = p_1 + p_2, \ldots, y_n = \sum_1^n y_n, \ldots$.

Generate a uniform PRN ξ.

Find the i th y interval such that $y_{i-1} \leq \xi < y_i$ and select $X = x_i$ as the sampled value. Let $P(X)$ be the corresponding probability.

Since ξ is uniformly random,

$$P(X) = P(y_{i-1} \leq \xi < y_i) = y_i - y_{i-1} = p_i \tag{4.13}$$

4.4.2.3 Sampling from a Generic Continuous Distribution

Consider a *generic continuous* pdf $f(x)$.

Integrate the probability density function $f(x)$ analytically or numerically and normalize it to 1 to obtain the normalized cumulative distribution:

$$F(t) = \frac{\int_{x_{min}}^{t} f(x)\mathrm{d}x}{\int_{x_{min}}^{x_{max}} f(x)\mathrm{d}x}. \tag{4.14}$$

Generate a uniform PRN ξ.

Get a sample of $f(x)$ by solving $\xi = F(X)$; that is, find the inverse value $X = F^{-1}(\xi)$ analytically or most often numerically by using look-up tables and interpolation. Let $P(X)$ be the correspondent probability.

Since ξ is uniformly random,

$$P(a \leq X < b) = P[F(a) \leq \xi < F(b)] = F(b) - F(a) = \int_a^b f(x)\mathrm{d}x. \tag{4.15}$$

Sampling from an exponential distribution is described in the following as a typical application. This is frequently needed in particle transport to find the point of the next interaction or the distance to decay.

$$f'(x) = \frac{e^{-x/\lambda}}{\lambda}, \quad x \in [0, \infty). \tag{4.16}$$

The cumulative distribution is

$$F'(t) = \int_0^t \frac{e^{-x/\lambda}}{\lambda} dx = 1 - e^{-t/\lambda}. \tag{4.17}$$

Generate a uniform PRN ξ. We will have $\xi = F'(X) = 1 - e^{-X/\lambda}$, and X can be sampled by inverting: $X = -\lambda \ln(1 - \xi)$.

4.4.2.4 The Rejection Technique

Most pdfs cannot be easily sampled by integration and inversion. An effective technique, the rejection technique, can be schematized as follows:

1. Let $f'(x)$ be a pdf that we want to sample.
2. Let $g'(x)$ be another pdf that can be sampled, such that $Cg'(x) \geq f'(x)$ for all $x \in [x_{min}, x_{max}]$.
3. Sample X from $g'(x)$ (i.e., generating a uniform PRN ξ_1 to sample like in the previous example).
4. Generate a second PRN ξ_2.
5. Accept X as a sample of $f'(x)$ if $\xi_2 < f'(X)/[Cg'(X)]$; otherwise, resample X and ξ_2 and repeat.

The probability of X to be sampled from $g'(x)$ is $g'(X)$, while the probability that it passes the test is $f'(X)/[Cg'(X)]$; therefore, the probability to have X sampled and accepted is the product of probabilities $g'(X)f'(X)/[Cg'(X)] = f'(X)/C$.

Since $f'(x)$ is normalized, the overall efficiency (probability accepted/rejected) is given by

$$\varepsilon = \int \frac{f'(x)}{C} dx = \frac{1}{C}. \tag{4.18}$$

To prove that the sampling is unbiased, that is, that X is a correct sample from $f'(x)$, we observe that the probability $P(X)dx$ of sampling X is given by

$$P(X)dx = \frac{1}{\varepsilon} g'(X)dx \frac{f'(X)}{Cg'(X)} = f'(X)dx. \tag{4.19}$$

The $g'(x)$ distribution can be as simple as a uniform (rectangular) distribution or as a normalized sum of uniform distributions (i.e., a piecewise constant function). In most cases, or when the x range is infinite, more complex choices are required.

Example 1. Let be $f'(x) = \frac{3}{8}(1 + x^2), x \in [-1, 1]$.

We choose $Cg'(x) = \max(f'(x)) = \frac{3}{4}, g'(x) = \frac{1}{2}$, and $C = \frac{3}{2}$.

Generate two uniform PRNs $\xi_1, \xi_2 \in [0, 1)$. Sample $X = 2\xi_1 - 1$.

Test: If $\frac{3}{8}(1 + X^2) / [Cg'(x)] = (1 + X^2)/2 > \xi_2$, accept X; otherwise, resample ξ_1 and ξ_2.

4.4.2.5 Other Sampling Techniques

Most discrete distributions, for instance, the Poisson distribution, cannot be expressed by a simple enumeration of probabilities. In a similar way, many continuous distributions cannot be integrated and inverted analytically and cannot be sampled easily by a rejection technique. Although inversion of the cumulative distribution (discrete or continuous) and rejection are the two basic sampling techniques, many other schemes have been found. Others are sometimes preferred because they are faster or easier to implement; for instance, a pdf $f(x) = x^n$ can be easily sampled by taking the largest of $n + 1$ random numbers. A comprehensive collection of such "recipes" can be found in the *Monte Carlo Sampler* by Everett and Cashwell [10].

4.5 Particle Transport Monte Carlo

A typical Monte Carlo particle transport code works as follows: Each particle is followed on its path through matter. At each step, the occurrence and outcome of interactions are decided by random selection from the appropriate probability distributions. All the secondaries issued from the same primary are stored in a "stack" or "bank" and are transported before a new history is started.

Most Monte Carlo transport codes are based on a number of assumptions, not always explicitly stated, which may limit their field of application or require some approximation.

- Media and geometry are generally supposed to be *static, homogeneous, isotropic,* and *amorphous*. A static geometry makes it difficult to handle problems with moving targets, although some attempts have been made successfully [11, 12] to overcome this limitation.
- Particle transport is considered to be a *Markovian process*; that is, the fate of a particle depends only on its actual present properties and not on previous or future events or histories.
- Particles *do not interact with each other*. This is a reasonable assumption in normal situations but not in extremely intense radiation fields as could be met in a plasma or inside a star. Also, effects like space charge and some rare radiation interactions occurring at high energies cannot be simulated, such as the Chudakov effect [13].

- Particles *interact with individual electrons, atoms, nuclei*, and *molecules*. This limitation does not allow simulation of coherent effects such as channeling in crystals or X-ray mirror reflection, which are important in synchrotron radiation optics.
- *Material properties are not affected by particle reactions.* Most Monte Carlo programs require a static material definition, but of course in these conditions it is not possible to handle problems such as burnup in nuclear reactors, where an intense neutron field modifies the isotope composition of a material.

The accuracy and reliability of a Monte Carlo depend on the models or data on which the probability density functions are based. The statistical accuracy of the results depends on the number of "histories." Statistical convergence can be accelerated by "biasing" techniques.

4.5.1 Monte Carlo Categories

4.5.1.1 Microscopic Analog Monte Carlo

The microscopic analog Monte Carlo type uses theoretical models to describe physical processes whenever it is possible, samples from actual physical phase space distributions, and predicts average quantities as well as higher moments. Codes belonging to this family, which preserve correlations and reproduce fluctuations in the best way allowed by the physics models used, can really be considered simulation codes. They are reasonably safe and generally do not require particular precautions by the user but are often inefficient and converge slowly. Since this type of Monte Carlo samples from the actual modeled physical distributions, it can fail to predict contributions due to rare but important events.

4.5.1.2 Macroscopic Monte Carlo

The macroscopic Monte Carlo type of particle transport, called also parameterized Monte Carlo, instead of simulating interactions in detail, uses parameterizations of the reaction product distributions obtained from fits to data and extrapolations. It is faster than analog microscopic simulations, especially when there are complex reactions, and can be more accurate if the theory contains uncertainties or approximations. It reproduces the single probability density functions but not the correlations among the products of the interactions. Of course, macroscopic Monte Carlo cannot be extended outside the range of the data used for the parameterizations.

4.5.1.3 Model-Based and Table-Based Codes

Originally, before 1980, there were three different types of Monte Carlo radiation transport codes:

- Low-energy neutron–photon codes, developed mainly for nuclear reactor problems (criticality and shielding).
- High-energy hadron codes, designed for accelerator shielding or for high-energy physics.
- Electron–photon codes.

Presently, some codes in the first category (i.e., MCNP [14]) and some belonging to the second (e.g., FLUKA [15]) have been extended to cover the whole energy range and to transport a large number of particles, including electrons.

4.5.1.4 Biased Monte Carlo

In biased Monte Carlo, one can sample from artificial distributions, applying a weight to the particles to correct for the bias. This is the mathematical equivalent of solving an integral by a change of the variable. This form of Monte Carlo predicts average quantities but not the higher moments; on the contrary, its goal is to minimize the second moment. Indeed, getting the same mean with smaller variance results in faster convergence. Biasing, more properly called variance reduction, techniques allow sometimes to obtain acceptable statistics when an analog Monte Carlo would take years of CPU (central processing unit) time to converge but cannot reproduce correlations and fluctuations. It must also be pointed out that biased Monte Carlo makes only privileged observables converge faster (some regions of phase space are sampled more at the expense of others).

4.6 Geometry

The algorithms to build a geometry and to track particles inside it differ from code to code. In some codes, the geometry is built from basic solids, in others from surfaces, and in some others from both. The user describes the geometry in question by input "cards" (text lines) or, in some codes, by user-written routines. In some cases, it is possible to define repeated structures obtained by translations or rotations of basic prototypes. Complex geometries such as that of a human body can be described in some Monte Carlo codes by "voxels" (small elementary parallelepipeds); this feature is typically used to import computed tomographic scans.

4.7 Monte Carlo Events

Particle histories are sequences of various events or processes. Some of them are physical (e.g., a particle interaction with an atom or with a nucleus); others are related to the transport or, better, to the geometry description, for instance, crossing a boundary between two materials. Events belong to two categories: discrete (or point-like) and continuous. Actual physical events are essentially discrete, but for convenience certain sequences of many similar, microscopic events are described as a continuous macroscopic one, sampled from a suitable distribution.

4.7.1 Discrete Processes

Discrete physics processes include atomic interactions by photons (Compton, photoelectric effect, pair production, coherent scattering) and by charged particles (bremsstrahlung, δ-ray emission, large-angle Coulomb scattering). Each of these processes is sampled only when the energy of a particle is higher than given thresholds, whether built in or set by the user. Nuclear interactions include absorption and nuclear scattering (elastic and nonelastic). Decays are also a type of discrete physics processes.

Point-like transport processes are boundary crossing and escape from the problem geometry.

4.7.2 Continuous Processes

Charged particles loose energy and change direction as a result of thousands of discrete collisions with atomic electrons. To simulate in detail each collision would require prohibitive computer times, except at very low particle energies. Therefore, generally many discrete scatterings are replaced by a straight continuous step, and the corresponding energy losses and changes of direction are "condensed" into a sum of losses (stopping power) and an overall scattering angle. This approach, common to most Monte Carlo particle transport programs, is known as the condensed history technique [16]. Some programs, however, provide a user option to simulate in detail the single scatterings in particular situations (very low energies, boundary crossing, conditions required by a multiple-scattering theory not satisfied).

Ionization energy losses belong to this type of continuous physical processes; all losses lower than a preset threshold are continuously distributed along a particle step. Any loss larger than the threshold is simulated as a discrete energy imparted to an electron, which is then transported separately (δ-ray). Energy loss fluctuations can be simulated for the losses below the threshold.

Multiple Coulomb scattering is another aspect of the same continuous process; a deflection angle, sampled from a theoretical distribution, is applied to each particle step. Some corrections are needed to account for the ratio between the length of the straight step and the actual path length (path length correction, PLC), for the lateral displacement, and so on [17].

When magnetic or electric fields are present, charged particle steps must be subdivided into smaller steps to follow the curvature of the trajectory while keeping dE/dx and the multiple scattering approximations.

4.7.3   Thresholds and Cutoffs

Transport and production thresholds are needed because of the limits of validity of the physics models and to reduce computer time.

4.7.3.1   Transport Thresholds

When the energy of a particle becomes lower than a specified transport threshold, the transport of that particle is terminated, and its remaining energy is deposited at that point or, better, in the case of a charged particle it is distributed along its residual range. The user's choice of a threshold will depend on the "granularity" of the geometry or of the scoring mesh and on the interest in a given region. To reproduce faithfully electronic equilibrium, neighboring regions should have the same discrete electron (δ-rays) energy threshold and *not* the same range threshold. Because photons travel longer paths than electrons of the same energy, photon thresholds should be lower than electron thresholds.

4.7.3.2   Production Thresholds

Energy thresholds need to be set also for explicit production of secondaries by photons and electrons. A δ-ray threshold sets the limit between discrete and continuous ionization energy losses. In a similar way, an energy threshold can be set for explicit production of bremsstrahlung photons; below that threshold, the electron radiative energy loss will be included in the continuous stopping power.

4.8   Biasing

Biased sampling, namely, sampling from nonnatural probability distributions, accelerates statistical convergence. This can be used for two different, and often complementary aspects: reducing the variance of a detector score for the same computer time or reducing the computer time needed to attain the same variance. To evaluate the effectiveness of a biasing technique, it is customary to define a figure of merit for an estimator called "computer cost" [18]:

$$F = \sigma^2 t \tag{4.20}$$

where σ^2 is the variance of a detector estimator, and t is the CPU time per primary particle.

Some biasing techniques aim to reduce σ^2, others to reduce t, but all are generally referred to as variance reduction techniques. The variance σ^2 converges like $1/N$ (where N = number of particle histories), while the computer time t is obviously proportional to N; therefore, minimizing $\sigma^2 t$ means reducing σ^2 at a faster rate than t increases or vice versa. The choice depends on the problems, and sometimes the combination of several techniques is the most effective.

4.8.1 The Two Basic Rules of Biasing

In the Boltzmann equation, there are two main ingredients: the particle angular flux $\Psi(x)$ (where x stands for all phase space coordinates) and various operators acting on $\Psi(x) [S(x), \beta(x), \Sigma_s]$. The operators are based on pdfs. The angular flux and the pdfs can both be biased, assigning to each particle a statistical weight.

The particle angular flux $\Psi(x)$ can be replaced by a fictitious angular flux $\Psi'(x)$. The particle weight $w(x)$ must be replaced by a new weight $w'(x)$ such that $w(x)\Psi(x) = w'(x)\Psi'(x)$. For instance, one particle can be replaced by two identical particles with half its weight and with the same position, energy, and direction.

First rule of biasing:

Weight $\times$ angular flux must be conserved

The pdf-based operators appearing in the Boltzmann equation are

- Source S: space–energy–angular probability distribution.
- $e^{-\beta}$: probability distribution of distance to interaction.
- Scattering cross section Σ_s: double-differential probability distribution in energy and angle.

A pdf $P(x)$ can be replaced by a fictitious pdf $P'(x)$. The particle weight $w(x)$ must be replaced by a new weight $w'(x)$ such that $w(x)P(x) = w'(x)P'(x)$.

Second rule of biasing:

Weight $\times$ probability must be conserved

In the following, a few common biasing techniques are briefly described.

4.8.2 Importance Biasing

Importance biasing acts on the particle fluence. The goal is to achieve a situation in which the number of contributions to fluence sampled in a given phase space region is proportional to its share of the score by a detector of interest. The most common form of importance biasing combines two techniques: surface splitting (also called geometry splitting), which reduces σ^2 but increases t, and Russian roulette (RR), which does the opposite. This kind of importance biasing is the simplest and easiest to use of all variance reduction techniques. If used alone, it is also safe since it introduces only small or zero weight fluctuations.

The user assigns a relative importance to each geometry region or cell (the actual absolute value does not matter) based on one or both of the following criteria:

1. The expected fluence attenuation in that region with respect to other regions.
2. The expected probability of contribution to score by particles entering that region.

Importance biasing is commonly used to keep the particle population constant, compensating for attenuation due to absorption or distance (first criterion), or to reduce sampling in space regions that are not likely to contribute to the result (second criterion).

4.8.2.1 Surface Splitting

If a particle crosses the boundary of a region, coming from a region of importance I_1 and entering a region of higher importance $I_2 > I_1$, the particle is replaced on average by $n = I_2/I_1$ identical particles with the same phase space coordinates (type, position, energy, direction). The weight of each "daughter" is that of the parent particle multiplied by $1/n = I_1/I_2$.

4.8.2.2 Russian Roulette

Russian Roulette (RR) acts in the opposite direction of splitting. If a particle crosses a boundary of importance I_1 to one of lower importance $I_2 < I_1$, the particle is submitted to a random survival test: With probability I_2/I_1, the particle survives with its weight increased by a factor I_1/I_2, and with probability $(1 - I_2/I_1)$ the particle is killed (its weight is set to zero).

4.8.3 Weight Window

The weight window technique is also a combination of splitting and RR, but it is based on the absolute value of the weight of each individual particle rather than on relative region importance. The user sets an upper and a lower weight limit, generally as a function of the region, energy, and particle. Particles having a weight larger than the upper limit are split so that the weight of their daughters will get a value between the limits (i.e., it will be brought "inside the window"). Particles having a weight smaller than the lower limit are submitted to RR: They are killed or have their weight increased to bring them "inside the window," depending on a random choice. The new weight is calculated so that average weight is conserved. The weight window is a more powerful biasing tool than simple RR or splitting based on region importance, but it requires more experience and patience to set up correctly since the expected absolute weights are not known a priori. It has often been said that "it is more an art than a science," but a "weight window generator" has been implemented in the MCNP program [14], which helps the user in this task [19], although it requires some experience to generate useful results and entails some additional CPU cost.

The effect of a weight window can be understood as follows: Killing a particle with a very low weight (with respect to the average for a given phase space region) decreases the computer time per history but has little effect on the score (and therefore on the variance of the score). On the other hand, splitting a particle with a large weight increases the computer time per history in proportion to the number of additional particles to be tracked, but at the same time reduces the variance by avoiding large fluctuations in the contributions to scoring. The global effect is to reduce the figure of merit $\sigma^2 t$.

4.8.4 Time Reduction: Leading Particle Biasing

Leading particle biasing (LPB) is used to avoid the geometrical increase with energy of the number of particles in a electromagnetic and hadronic shower [20]. In every electromagnetic interaction, two particles are present in the final state (at least for pair production and bremsstrahlung in the approximations made by most Monte Carlo codes). With LPB, only one of the two particles is randomly kept, and its weight is adjusted to conserve weight × probability. The most energetic particle is kept with higher probability, proportional to its energy, as the one that is more efficient in propagating the shower, and its weight is adjusted to account for the bias.

Let us assume that a particle with weight w produces two particles P_1 and P_2 with energy respectively E_1 and E_2, with $E_1 > E_2$. Their total unbiased weight, inherited from the parent, will be $2w$.

- Particle P_1, chosen with probability $E_1/(E_1 + E_2)$, will be assigned a weight $w \times (E_1 + E_2)/E_1$.
- Particle P_2, with probability $1 - E_1/(E_1 + E_2) = E_2/(E_1 + E_2)$, will have a weight $w \times (E_1 + E_2)/E_2$.

Weight conservation is ensured:

$$w = \frac{E_1}{E_1 + E_2}\frac{E_1 + E_2}{E_1} + w\frac{E_2}{E_1 + E_2}\frac{E_1 + E_2}{E_2} = 2w. \qquad (4.21)$$

LPB is effective at reducing the CPU time per history t but increases the score variance σ^2 by introducing large weight fluctuations since a few low-energy particles end up carrying a high weight, while many energetic particles will have a small weight.

Therefore, LPB should be always backed up by a weight window. A similar scheme can be applied to hadronic nuclear interactions as well, by which several particles are routinely produced. In this more general case, a predetermined fraction of the particles generated is kept rather than keeping one particle only. They are selected using a probability proportional to their energy, and their weight is adjusted accordingly.

There is an important warning: With LPB, as with many biasing schemes, the energy is conserved only on average.

4.8.5   Nonanalog Absorption

Nonanalog absorption, also called survival biasing, is usually applied to low-energy neutron transport. There are three possibilities to handle neutron scattering and absorption (we indicate with σ_s the scattering cross section and with σ_t the total cross section):

1. Analog: At each collision, scattering and absorption are sampled with the actual physical probability: σ_s/σ_t and $(1-\sigma_s/\sigma_t)$, respectively. If scattering is selected, the weight is unchanged; if absorption, the weight becomes zero.
2. Biased without absorption: systematic survival, weight reduced by a factor σ_s/σ_t.
3. Biased with user-defined absorption probability u. Scattering has probability $1-u$. The weight is reduced by a factor $\frac{\sigma_s}{\sigma_t(1-u)}$.

Possibility 2 is a special case of the third possibility with $u = 0$. The third option is available in the FLUKA code [15]. Note the exchanges between probabilities and weights, such that the product Probability $\times$ Weight will remain unchanged.

Biased survival without absorption makes a neutron remain alive forever, except for escaping from the geometry. After many collisions, the neutron will have a very small weight. In codes where this option is chosen, RR or a weight window technique are needed at each collision to avoid tracking neutrons with a weight too small to make a significant contribution to the score.

4.8.6   Biasing Mean Free Paths

4.8.6.1   Decay Length

The mean life or the average decay length of unstable particles can be artificially shortened to increase the generation rate of decay products. This technique is typically used to enhance statistics of muon or neutrino production. The kinematics of decay (decay angle) can also be biased to enhance sampling in a preferred direction.

4.8.6.2 Interaction Length

In a similar way, the hadron or photon mean free path for nuclear inelastic interactions can be artificially decreased by a predefined particle- or material-dependent factor. This option is useful, for instance, to increase the probability of beam interaction in a thin target or in a material of low density.

Mean free path biasing, for decay or inelastic interactions, can be done in two ways:

1. The parent particle does not disappear; the weight of the particle and of its secondaries is modified to take into account the ratio between biased and physical survival probability.
2. The survival of the parent particle is decided by RR by comparing a random number to the ratio between biased and physical survival probability; if the particle survives, its weight is not modified. The secondaries are created in any case, whether the parent survives or not, and their weight is modified to take into account the ratio between biased and physical survival probability.

Interaction length biasing is also necessary to sample photonuclear reactions with acceptable statistics since the photonuclear cross sections are much smaller than those for electromagnetic processes.

4.8.6.3 Sampling from a Biased Distribution

The sampling technique based on the cumulative pdf can be extended to bias the sampling probability in different parts of the interval (importance sampling).

Assume that $f(x)$ is replaced by $g(x) = f(x)h(x)$, where $h(x)$ is any appropriate weight function. We normalize $f(x)$ and $g(x)$:

$$f'(x) = \frac{f(x)}{\int_{x_{min}}^{x_{max}} f(x)\mathrm{d}x} = \frac{f(x)}{A} \tag{4.22}$$

$$g'(x) = \frac{g(x)}{\int_{x_{min}}^{x_{max}} g(x)\mathrm{d}x} = \frac{g(x)}{B} \tag{4.23}$$

and consider the biased cumulative normalized distribution $G(x)$:

$$G(t) = \frac{\int_{x_{min}}^{t} g(x)\mathrm{d}x}{B} \tag{4.24}$$

Now, let us sample from the biased cumulative normalized distribution G instead of the original unbiased F; let us take a uniform random number ξ and get the sampled value X by inverting $G(x)$:

$$X = G^{-1}(\xi) \tag{4.25}$$

The particle weight must be multiplied by the ratio of the unbiased to the biased normalized pdf at $x = X$:

$$w' = w\frac{f(X)}{g(X)} = w\frac{B}{A}\frac{1}{h(X)} \tag{4.26}$$

A special case occurs when the biasing function chosen is the inverse of the unbiased pdf:

$$h(x) = \frac{1}{f(x)} \quad g(x) = f(x)h(x) = 1 \tag{4.27}$$

$$\begin{aligned} G(x) &= \frac{\int_{x_{\min}}^{x} g(x)\mathrm{d}x}{\int_{x_{\min}}^{x_{\max}} g(x)\mathrm{d}x} = \frac{x - x_{\min}}{B} \\ &= \frac{x - x_{\min}}{x_{\max} - x_{\min}}. \end{aligned} \tag{4.28}$$

In this case, $X = G^{-1}(\xi) = x_{\min} + \xi(x_{\max} - x_{\min})$ and the weight of the sampled particle is multiplied by

$$\frac{B}{A}\frac{1}{h(X)} = \frac{x_{\max} - x_{\min}}{A}f(X). \tag{4.29}$$

Because X is sampled with the same probability over all possible values of x, independently of the value $f(X)$ of the function, this technique is used to ensure that sampling is done uniformly over the whole interval, even though $f(x)$ might have small values in some x range. For instance, it may be important to avoid undersampling in the high-energy tail of a spectrum, steeply falling with the energy but more penetrating at high energies, such as that of cosmic rays or synchrotron radiation.

4.9 Monte Carlo Results

It is often said that a Monte Carlo calculation is a "mathematical experiment" [4,21]. Each aspect of a real experiment indeed has its Monte Carlo equivalent:

Experimental technique $\Rightarrow$ Estimator
Instrument $\Rightarrow$ Detector
Measurement $\Rightarrow$ Score or tally
Result of an experiment $\Rightarrow$ Monte Carlo result

Just like a real measurement, a score is obtained by sampling from a statistical distribution. As an experimental result consists of an average of measurement values, a statistical error, and a systematic error, a Monte Carlo result is an average of scores, a statistical error, and a systematic error generally unknown.

4.9.1   Estimators

There are often several different techniques to measure the same physical quantity; in the same way, the same quantity can be calculated with different kinds of estimators.

4.9.1.1   Estimator Types

There are various types of estimators, depending on the quantity to be estimated and on its topology (phase space region over which the quantity is integrated).

- The *boundary crossing* estimator is used to estimate the fluence or the current of particles at a physical boundary between two space regions. Possible results are mono- or multidifferential fluence spectra as a function of energy, angle, particle type, and so on.
- The *track length* estimator calculates the fluence of particles in a region of real space. The results are fluence spectra as a function of particle energy and type based on their path lengths within the region volume.
- A *pulse height* estimator is used to simulate the response of a spectrometer (i.e., a Ge detector). The quantity estimated is the energy deposited in a region of real space, and the result is the spectrum of deposited energy within the region volume.
- Scalar integral estimators are used to predict scalar quantities, or their densities, such as deposited energy, inelastic interactions ("stars"), induced activity, and so on in a region of real space.
- A *mesh* (or *binning*) estimator is a special case of scalar estimator, providing a two- or three-dimensional space distribution of a scalar quantity (scalar fluence, energy deposition, star density) over a regular subdivision of a portion of real space in subvolumes, generally independent from the tracking geometry.

4.9.2   Detectors

While an estimator is a technique to "measure" a certain quantity, a detector is an instantiation (a concrete application) of an estimator in a particular region of phase space. For instance, a track length estimator of fluence can be concretely applied as a particular detector consisting of a sphere of given radius centered at specified coordinates. Often, a Monte Carlo user wants to get a result at one or more detectors and has no interest in what happens elsewhere. In this case, biasing can be used to accelerate convergence in the neighborhood of a detector at the expense of other parts of phase space (but if biasing is done correctly, for $N \rightarrow \infty$, the integration will converge to the true value everywhere, although at a different speed). Thanks to modern fast computers, however, it is often possible to obtain a good result with a multiple detector (a mesh detector) covering a large region of real space. This kind of detector can be useful for identifying shielding weaknesses in unexpected places, as revealed, for instance, by color plots.

4.9.3   Statistical Errors

The variance of the mean of an estimated quantity x (e.g., fluence), calculated in N batches, is given by

$$\sigma^2_{<x>} = \frac{1}{N-1} \left[\frac{\sum_1^N n_i x_i^2}{n} - \left(\frac{\sum_1^N n_i x_i}{n} \right)^2 \right] \qquad (4.30)$$

where n_i is the number of histories in the ith batch, $n = \Sigma n_i$ = total number of histories in the N batches, and x_i is the average of x in the ith batch. The variance can be calculated for single histories (in the limit $N = n$, $n_i = 1$) or for batches of several histories each (not necessarily the same identical number).

The distribution of scoring contributions by single histories can be asymmetric since many histories contribute little or zero. Scoring distributions from batches tend to be Gaussian for $N \rightarrow \infty$, provided $\sigma^2 \neq \infty$ (central limit theorem).

The standard deviation of an estimator calculated from batches or from single histories is an estimate of the standard deviation of the actual distribution ("error of the mean"). The precision of such an estimate depends on the type of estimator and on the particular problem, but the sample average converges to the actual distribution average for $N \rightarrow \infty$.

4.9.3.1   Effect of Sampling Inefficiency on Statistical Errors

The following table, adapted from the MCNP manual [14], suggests that the actual meaning of a calculated statistical error may be different in Monte Carlo from that expected based on the statistics of experiments:

Relative error	Quality of tally
50–100%	Garbage
20–50%	Factor of a few
10–20%	Questionable
<10%	Generally reliable

Why does a 30% calculated error mean in fact an uncertainty of a "factor of a few"? This is because the actual error corresponds to the sum (in quadrature) of two uncertainties: one due to the fraction of histories that give a zero contribution and one that reflects the spread of the nonzero contributions. Often, this results in a distribution of "hits" in the detector of interest is far from Gaussian. Furthermore, by definition, the error computed according to 4.30 can never exceeds 100% if at least two hits have been recorded. Finally, in many situations the score is dominated by few, large-weight, events, which can be undersampled and result in a sudden jump in the result as soon as one of them is recorded. The MCNP guideline is empirically based on experience, not on a mathematical proof, but it has been generally confirmed also in other codes.

4.9.4   Other Errors

Just as in experimental measurements, in addition to statistical errors Monte Carlo results can be affected by systematic errors and in some unfortunate cases even by mistakes. Systematic errors arise from code weaknesses (e.g., in the physics models), lack of information (e.g., in material composition, particularly when trace elements are an issue), and problem simplification (e.g., in the geometry description).

4.10   Quality Assurance

Many enforcing authorities require that quality assurance (QA) must be applied to design procedures. But, imposing it on Monte Carlo calculations can only be done in a limited way. Monte Carlo is not a "black box," and different users are likely to choose different approaches, equally valid, to the same problem. Among other difficulties, a strict QA would probably require forbidding to write user code and would be incompatible with the use of most variance reduction techniques ("they are more an art than a science," as reminded here). However, QA can be required for keeping proper documentation about code version, input, outputs, applied biasing, possible user code, assumptions, normalization, and so on. Other recommended QA features are that critical calculations have to be submitted to peer reviews and to audits made by independent experts.

References

1. Kalos, M.H. and Whitlock, P.A. *Monte Carlo Methods*, 2nd edition, Wiley-VCH, Berlin (2008)
2. Lux, I. and Koblinger, L. *Monte Carlo Particle Transport Methods: Neutron and Photon Calculations*, CRC Press, Boca Raton, FL (1990)

3. Carter, L.L. and Cashwell, E.D. *Particle-Transport Simulation with the Monte Carlo Method*, ERDA Critical Review Series, National Technical Information Service, Springfield, MA (1975)
4. Hammersley, G.M. and Handscomb, D.C. *Monte Carlo Methods*, Wiley, New York (1964)
5. Spanier, J. and Gelbard, E.M. *Monte Carlo Principles and Neutron Transport Problems*, Addison-Wesley, Reading, MA (1969)
6. Pólya, G. *Über den zentralen Grenzwertsatz der Wahrscheinlichkeitsrechnung und das Momentenproblem* (in German) Math. Z. **8**, 171–181 (1920)
7. Dupree, S.A. and Fraley, K. *A Monte Carlo Primer*, Kluwer/Plenum, New York (2002)
8. Marsaglia, G. and Tsang, W.W. *The 64-bit universal RNG*, Stat. Probabil. Lett. **66**, 183–187 (2004)
9. Matsumoto, M. and Nishimura, T. *Mersenne twister: a 623-dimensionally equidistributed uniform pseudo-random number generator*, ACM Trans. Model. Comput. S. **8**, 3–30 (1998)
10. Everett, C.J. and Cashwell, E.D. *A Third Monte Carlo Sampler*, Los Alamos Report LA-9721-MS (1983)
11. Biaggi, M., Ballarini, F., Burkard, W., Egger, E., Ferrari, A. and Ottolenghi, A. *Physical and bio-physical characteristics of a fully modulated 72 Me V therapeutic proton beam: model predictions and experimental data*, Nucl. Instrum. Meth. **B159**, 89–100 (1999)
12. Paganetti, H., Jiang, H. and Trofimov, A. *4D Monte Carlo simulation of proton beam scanning: modelling of variations in time and space to study the interplay between scanning pattern and time-dependent patient geometry*, Phys. Med. Biol. **50**, 983–990 (2005)
13. Chudakov, A.E., Izv. Akad. Nauk SSSR, Ser. Fiz. **19**, 650 (1955)
14. X-5 Monte Carlo Team *MCNP: a general Monte Carlo N-particle transport code*, Los Alamos Report No. LA-UR-03–1987 (2003)
15. Battistoni, G., Muraro, S., Sala, P.R., Cerutti, F., Ferrari, A., Roesler, S., Fassò, A. and Ranft, J. *The FLUKA code: Description and benchmarking*, Proceedings of the Hadronic Shower Simulation Workshop 2006, Fermilab 6–8 September 2006, M. Albrow, R. Raja eds., AIP Conference Proceeding 896, 31–49; (2007);Ferrari, A., Sala, P.R., Fassò, A. and Ranft, J. FLUKA: a multi-particle transport code, CERN-2005–10, INFN/TC_05/11, SLAC–R-773 (2005)
16. Berger, M.J. Monte Carlo calculation of the penetration and diffusion of fast charged particles, in *Methods in Computational Physics*, ed. B. Alder, S. Fernbach and M. Rotenberg (Academic Press, New York), Vol. 1, pp. 135–215 (1963)
17. Ferrari, A., Sala, P.R., Guaraldi, R., and Padoani, F. *An improved multiple scattering model for charged particle transport*, Nucl. Instrum. Meth. **B71**, 412–426 (1992)
18. Juzaitis, R.J. *Minimizing the cost of splitting in Monte Carlo radiation transport simulation*, Thesis, Los Alamos Report LA-8546-T (1980)
19. Booth, T.E. and Hendricks, J.S. *Importance estimation in forward Monte Carlo calculations*, Nucl. Technol. Fusion **5**, 90 (1984)
20. Van Ginneken, A. *AEGIS – a program to calculate the average behavior of electromagnetic showers*, Fermilab-FN-309 (1978)
21. Galison, P. Computer simulations and the trading zone, in *The Disunity of Science: Boundaries, Contexts, and Power*, ed. P. Galison and P.J. Stump, Stanford University Press, Stanford (1996)

Part III
Radiation Sources and
Radiopharmaceutical Productions

Chapter 5
Sealed Radionuclide and X-Ray Sources in Nuclear Medicine

Sören Mattsson and Arne Skretting

5.1 Introduction

Apart from the radiopharmaceuticals that are administered to the patient as part of an examination or treatment [1], nuclear medicine uses different radiation sources, both as part of the examination procedures and for quality control of the equipment involved. In connection with the examination, radiation sources are used as markers and for measurements of transmission of radiation through the body as a basis for attenuation correction. The performance of activity meters, scintillation cameras, and positron emission tomographic (PET) scanners are all monitored by the use of solid radiation sources. It is preferable that these sources have long half-lives to simplify constancy checks and to avoid frequent costly replacements.

Today, an increasing number of nuclear medicine procedures are carried out using "hybrid imaging" [2]. Images that show different aspects of anatomy and physiology are acquired by different – mainly three-dimensional – techniques and shown on one screen with all image sets geometrically aligned and often displayed on top of one another. Thus, one finds combined single-photon emission computed tomography and computed tomography (SPECT/CT), PET/CT, SPECT/PET, and PET/MR (magnetic resonance). This development has brought CT and thereby X-ray sources into the practice of nuclear medicine. Here, CT images are used not only for identification of anatomical details but also as a basis for attenuation correction. Older techniques for attenuation correction (i.e., measurement of the transmission through the body using radiation from a radionuclide source) are now being increasingly substituted by CT scanners incorporated in the equipment.

S. Mattsson (✉)
Medical Radiation Physics, Department of Clinical Sciences Malmö, Lund University, Skåne University Hospital, 205 02 Malmö, Sweden
e-mail: soren.mattsson@med.lu.se

A. Skretting
Section of Diagnostic Physics, The Interventional Centre, Oslo University Hospital, Ullernchausseen 70, 0310 Oslo, Norway

M.C. Cantone and C. Hoeschen (eds.), *Radiation Physics for Nuclear Medicine*,
DOI 10.1007/978-3-642-11327-7_5, © Springer-Verlag Berlin Heidelberg 2011

5.2 Sources

5.2.1 Sealed Sources for Constancy Control of Activity Meters

Sources for calibration and constancy control of activity meters should have long
physical half-lives, a range of photon energies, and a range of activities and be
activity calibrated within ±5% or better [3] (Table 5.1).

"Mock sources" are sources (often long lived) used to simulate the real, clinically
used sources, such as mock ^{99m}Tc made from ^{57}Co or ^{141}Ce, mock ^{131}I of ^{137}Cs
or ^{133}Ba, mock ^{125}I of ^{129}I, and so on. ^{60}Co and ^{226}Ra are used for checking
the long-term stability of the activity meters and surface monitors. If there is no
calibration source, an approximate calibration can be obtained from information
about the specific gamma-ray dose constant Γ for the radionuclide in question by
interpolation between radionuclides of nearby photon energies (refer to Fig. 5.1) for
which calibrations exist.

Table 5.1 Sources used for constancy control of activity meters (and gamma counters for in vitro measurements)

Radionuclide	Photon energy (keV)	Half-life	Typical activity (MBq)
^{57}Co	122; 136	271 days	~40
^{141}Ce	145	32.5 days	~40
^{133}Ba	81; 356	10.7 years	~10
^{137}Cs	662	30 years	~4
^{129}I	30	1.6×10^{7} years	~4
^{60}Co	1,173; 1,332	5.27 years	~2
^{226}Ra (in equilibrium with its decay products)	186; 242; 295; 352; 609	1,600 years	~2

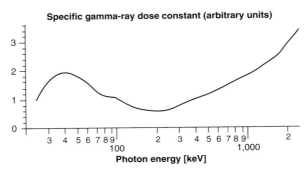

Fig. 5.1 Example of sensitivity variation with photon energy for a cylindrical activity meter
(erroneously often still called "dose calibrator")

5.2.2 Sealed Sources Used for Marking of Anatomical Locations in Images

Sealed sources of ^{57}Co are commonly used in pencils, rulers, or flexible lines to mark anatomical structures so that these can be shown superimposed on the diagnostic images. Such sources have an activity of 1–8 MBq at purchase and are used for a period of around 2 years. Another radionuclide that is used for this purpose is ^{241}Am (59 keV). The photon energies of these sources are different from that of the photons emitted from the actual radionuclide inside the body, and this means that the energy window has to be shifted for imaging the markers. It is therefore important that the positional energy linearity of the scintillation camera is acceptable. In some older models of cameras (working mainly with analog electronics), one could experience erroneous positioning of the marker image relative to the diagnostic image. As an example, photons from ^{131}I and ^{57}Co emitted from the same position on the camera surface would in case of such errors show up in different positions in the image.

5.2.3 Sources for the Control of Gamma Cameras

To regularly control uniformity, sensitivity, spatial resolution, spatial linearity, energy resolution, and "peaking" of gamma cameras, there is a need for a number of point, line, collimated line, planar, and flood sources. The performance parameters are checked without the collimator ("intrinsic" tests) and with the collimator mounted ("extrinsic" tests).

For uniformity *correction*, an image of a uniform source is obtained with a high number of counts. This procedure requires tenfold activity compared to that used for the regular flood source (uniformity) measurements.

The intrinsic uniformity is measured using a point source of around 20 MBq at a distance of about five times the field-of-view (FOV) diameter of the camera to ensure uniform irradiation of the detector. Intrinsic spatial resolution can be assessed with a four-quadrant bar phantom placed on the detector and a similar point source arrangement.

Measurements of extrinsic uniformity can be performed using either a solid disk of ^{57}Co ($T_{1/2} = 270$ days; 122 keV) with an activity of 400–600 MBq at the time of purchase or a refillable plastic source containing a mixture of 400–600 MBq ^{99m}Tc (^{99m}Tc : $T_{1/2} = 6$ h.; 140 keV) (or any other radionuclide of interest) and water. As with the intrinsic test, uniformity correction procedures require ten times the activity used for the regular acquisition with a uniform source.

Besides the difference in photon energy between ^{57}Co and ^{99m}Tc, ^{57}Co disk sources have two major drawbacks. They are expensive and need to be replaced every 1–2 years. In addition, new ^{57}Co sheet sources usually contain small amounts of ^{56}Co and ^{58}Co. These radionuclide contaminants have a shorter half-life

(70–80 days) than ^{57}Co and emit high-energy gamma rays (>500 keV). Visual inspection of the energy spectrum from a recently manufactured ^{57}Co sheet source will often show a significant amount of high-energy contamination. During the first few months of use, this contamination may adversely affect the results of the tests unless measurements of uniformity are performed with a medium- or high-energy collimator. The disadvantages of a ^{99m}Tc flood source are the time taken in preparing the source and the unavoidable radiation exposure to the personnel. In addition, refillable sources are prone to a number of other problems. These include geometric distortion of the walls of the container, presence of air bubbles inside the source (which can be avoided by bubble traps), poor mixing of the radionuclide within the source, and clumping or adhesion of the radionuclide to the walls of the plastic container.

For SPECT, the overall system performance may be evaluated using some of the commercially available liquid-fillable tomographic phantoms. These are usually circular phantoms containing a variety of rods or spheres (nonradioactive or radioactive inserts of different sizes) that can be filled with a mixture of water and ^{99m}Tc. Image quality is then evaluated by visual inspection. The spatial resolution can also be measured with a point source (made by immersing a small seed or gel ball of ^{99m}Tc), the image of which can be fitted to a Gaussian function to determine the full width at half maximum (FWHM).

The position of the rotational axis relative to the image matrix (center of rotation, COR) can be measured by acquisition of projection images of a point source of ^{99m}Tc viewing from angles that span 360°. Opposite views enable determination of the true position of the rotational axis and calculation of the offsets relative to the midpoint of the image matrix. These offsets are then stored and used for subsequent real-time corrections of the x (positional) signal before the actual count is assigned to a picture element.

5.2.4 Sources for Attenuation Correction of SPECT Measurements Through Transmission Measurements

5.2.4.1 Radionuclide Sources

Several SPECT systems currently use a transmission source of a radionuclide that is mounted opposite to the detector in different configurations [4]. An approach to attenuation correction for ^{99m}Tc imaging that has been used in the past used sealed line sources of ^{153}Gd ($T_{1/2}$ = 242 days) and ^{57}Co ($T_{1/2}$ = 270 days). These sources provide beams of around 100-keV photons (^{153}Gd photon energies of 97 keV in 29% of the decays and 103 keV in 21%; ^{57}Co photon energies 122 keV in 86% and 136 keV in 11%) and are scanned in the longitudinal direction at each step of the SPECT acquisition to provide transmission maps of the region under investigation (Fig. 5.2). The typical activity of a 25 cm long ^{153}Gd line source, encapsulated and collimated, is 30–800 MBq. For ^{201}Tl (70 keV in 59%, 80 keV

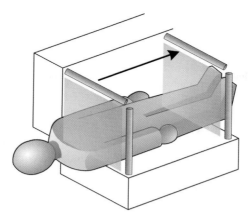

Fig. 5.2 Line sources are scanned over the gamma camera field of view to get transmission maps for different directions. These data are used for the reconstruction of μ maps

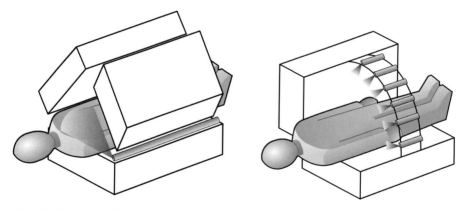

Fig. 5.3 Two alternative setups for the acquisition of transmission maps to be used for the reconstruction of μ maps

in 13%, 167 keV in 10%) imaging, ^{241}Am (59 keV in 36%) and ^{153}Gd (97 keV in 29%, 103 keV in 21%) are used in different configurations. Different other geometrical arrangements for measurement of transmitted radiation with sealed sources are illustrated in Fig. 5.3. Correction of images for attenuation effects is complicated by the broad range of tissue types (lung, soft tissue, muscle, and bone) that are present in the body volumes of interest. The goal of attenuation correction is to create a matrix of correction factors (refer to the next paragraph), with each matrix element corresponding to a voxel in the SPECT image. The basis for the calculation of such correction factors is an image set that contains in each voxel the value of the linear attenuation coefficient (a μ map) for photons from the actual radiopharmaceutical (see also the next section).

5.2.4.2 CT

A better alternative for attenuation correction in SPECT is the application of a built-in CT, which has the advantage of fast data collection. CT images are also useful for diagnostic purposes and for anatomic localization of a SPECT finding (by aligned superposition of the image sets, "fusion"). The CT attenuation map reconstructed from measurements with an effective CT photon energy of, say, 70 keV can easily be converted (using mass attenuation ratios for 70 and 141 keV in tissue) to form the μ maps corresponding to the SPECT photon energy of 141 keV. This combined modality is today the most sold alternative for attenuation correction.

Correction factors are derived for each voxel as the inverse of the mean attenuation factors for photons that are emitted from the actual voxel [4]. The mean attenuation factor can be determined from calculated attenuation factors (based on the μ map) along the paths in the directions of the different viewing angles. The correction factors are subsequently applied to the reconstructed SPECT images voxel by voxel.

Current technology using multidetector arrays (typically 64 detector rows) and helical (spiral) scanning permits acquisition of several hundred CT slices while the patient holds his or her breath. The X-ray tube may move around the patient in as little as 0.35 s, and the slice thickness can be as low as 0.6 mm.

The integration of an emission tomography system (SPECT or PET) and CT scanning with a common patient-handling system (bed) provides a significant advance in technology (Figs. 5.4 and 5.5). The two scanners are mechanically aligned. These combinations permit the acquisition of emission and transmission data sequentially in a single study with the patient in an ideally fixed position. Thus, the two data sets can be acquired in a registered format by appropriate calibrations, permitting the acquisition of corresponding slices from the two modalities.

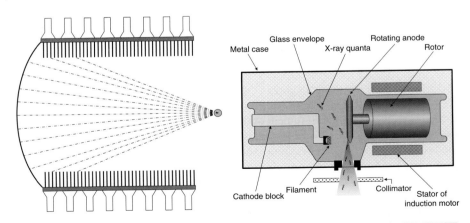

Fig. 5.4 A single-photon emission computed tomographic/computed tomographic (SPECT/CT) system with dual-head scintillation cameras and a low-output fan-beam CT scanner. A separate figure showing the principles of an X-ray tube is included to the *right*

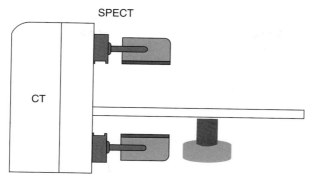

Fig. 5.5 An integrated dual-head scintillation camera single-photon emission computed tomographic (SPECT) system and a commercially available computed tomographic (CT) scanner

Since the CT images are acquired with smaller pixels than the emission data, it is necessary to decrease the resolution of the CT data (rebin to a smaller matrix size) to provide attenuation correction matrices for the emission study. Because of reduced noise in the rebinned images, adequate attenuation correction can be obtained even with a "low-dose" CT study.

Another advantage related to the high photon fluence rate of CT scanners is that attenuation measurements can be made in the presence of radionuclide distributions with negligible contributions from photons emitted by the radionuclides, which means that postinjection CT measurements can be performed. The use of CT also eliminates the need for additional hardware and transmission sources that often must be replaced on a regular basis.

5.2.4.3   Clinical SPECT/CT Systems

The clinical SPECT/CT systems currently used typically have dual-head scintillation cameras and a CT sharing a common imaging table. There are two approaches to clinical SPECT/CT applications. The first is the use of a low-output fan-beam CT scanner (Fig. 5.4) using an X-ray tube (with low tube current, ~2.5 mA). It can acquire four 5-mm anatomical slices in 13.6 s (e.g., General Electric Hawkeye). The CT images acquired with this system are not of sufficient quality for diagnostic procedures but are adequate for attenuation corrections and anatomical correlation with emission images. The slow scan speed is actually an advantage in regions where there is physiological motion since the CT image blurring from the motion is comparable to that of the emission scans, resulting in a good match in the fused images. The effective dose from this system is typically around 3 mSv, which is much higher than for applications using radioisotope transmission sources (0.1 mSv).

The second approach is to integrate commercially available CT scanners with dual-head scintillation cameras (Fig 5.5). Typically, tube currents of 20–500 mA are used, facilitating slice thicknesses of 0.6–10 mm and rotation times of 0.5–1.5 s.

These systems give approximately four to five times the patient radiation dose of that from a fan-beam CT system. Since the CT scanners in these systems are commercially available diagnostic systems, the images produced are of sufficient quality to be used for diagnostic purposes in addition to their obvious use for attenuation correction and anatomical correlation. Effective doses to the patient from examinations with these systems are on the order of 10–15 mSv when images of diagnostic quality are produced. Operated in a lower radiation dose mode (about 3 mSv) by reducing the X-ray tube current, these systems are still acceptable for attenuation correction and anatomical correlation applications.

5.2.5 Sources for Constancy Check and Attenuation Correction of PET Measurements Through Transmission Measurements

Different radiation sources are used in PET to calibrate detector efficiency (to calibrate for detector nonuniformities) and to assess attenuation in the patient. Examples of radionuclide sources are ^{68}Ge ($T_{1/2} = 271$ days), ^{137}Cs (30 years), and ^{22}Na (2.6 years). ^{68}Ge (or more correctly its daughter, ^{68}Ga) and ^{22}Na are positron emitters, and ^{137}Cs emits a single 662-keV photon. In PET/CT equipment, the CT is also used for attenuation correction.

All manufacturers have a standard procedure for the acquisition of PET "normalization" data.

Depending on the system and the acquisition mode (two and three dimensional), the normalization procedure can be accomplished using different sources and phantoms. The most widely used ones are as follows:

1. A rotating ^{68}Ge line source
2. A uniform cylindrical ^{68}Ge phantom centered horizontally and vertically in the FOV of the PET
3. A rotating ^{137}Cs point source

5.2.5.1 Clinical PET/CT Systems

Clinical PET/CT systems have CTs equipped with 4, 8, 16, 64, or even more detector rows and provide images of sufficient diagnostic quality to be used for diagnostic procedures. As with SPECT/CT systems, the CT scanners can be operated at reduced tube current if the scans are only to be used for attenuation correction and anatomical correlation.

For attenuation correction, earlier generations of PET scanners used rod sources of ^{68}Ge/^{68}Ga or ^{137}Cs that rotated slowly around the patient to measure transmission through the body at detector elements on the opposite side of the body. Matrices of attenuation coefficients were then calculated by basically the same algorithms that are used in modern CT scanners. Today, far more than 90% of the PET scanners that are purchased come with a CT scanner that is mounted on the same gantry as the PET scanner, and the patient can be transferred under computer control between PET scanner and a helical CT scanner. Again, corrections need to be carried out to take into account the differences in photon energies of 511 keV and the effective energy of the CT scanner (around 70–90 keV). Such corrections may be difficult to perform when iodine contrast is present in the body.

Attenuation correction is simpler in PET than in SPECT because both anni-hilation photons need to reach detectors on the opposite sides of the body. The probability of this outcome depends on the sum of attenuation coefficients along the line that connects the two detectors where the two photons hit. This means that the acquired activity projections (sinograms) can be corrected for attenuation before the reconstruction routine is run.

5.2.5.2   Radiation Dosimetry Considerations for SPECT/CT and PET/CT

The replacement of conventional transmission sources with CT increases the con-cern regarding radiation dose to the patient. In the past, the radiation dose from transmission sources such as ^{153}Gd used in SPECT and ^{68}Ge used in PET was gen-erally ignored due to the small contributions associated with their use in comparison to the dose from the radionuclide studies. For example, the effective dose for a PET transmission scan with ^{68}Ge sources is on the order of 0.1 mSv. However, when CT is used, the dose associated with this procedure is of such magnitude that it must be taken into consideration. The effective dose for a diagnostic CT of the chest, abdomen, pelvis is on the order of 10 mSv. For purposes of comparison, the effec-tive dose from a PET scan with 370 MBq of ^{18}FDG (^{18}F-fluoro-deoxy-glucose) is 11 mSv and from a SPECT scan with 900 MBq of ^{99m}Tc-labeled white cells is 18 mSv. If the CT scan is only to be used for attenuation correction, the tube current (mA) can be reduced by a factor of four or more, reducing the effective dose to no more than 3 mSv.

References

1. Holzwarth, U. Radiopharmaceutical production. (This volume).
2. Patton, J.A. History and principles of hybrid imaging. In: D. Delbeke, O. Israel (eds.), Hybrid PET/CT and SPECT/CT Imaging. A teaching file, Springer Science + Business Media, New York, pp 3–33 (2010).
3. IEC, International Electrotechnical Commission. *IEC TR 61948–4 Ed 1.0: Nuclear medicine instrumentation – Routine tests – Part 4: Radionuclide calibrators.* Geneva: IEC (2006).
4. Patton, J. A. and Turkington, T. G. SPECT/CT physical principles and attenuation correction. J Nucl Med Technol 36(1), 1–10 (2008).

Chapter 6
Radiopharmaceutical Production

Uwe Holzwarth

6.1 Introduction

Radiopharmaceuticals are medicinal formulations containing one or more radionuclides. The emissions of the radionuclide are used either in diagnostics to trace and
visualise the biodistribution of a substance or in therapy to deliver a high radiation dose to a target tissue. The particularity of radiopharmaceuticals consists of
their capability to retrieve information on a molecular level and to address systems with very low densities of receptor molecules in vivo and in a *noninvasive*
way [1, 2].

In diagnostic applications, a tiny concentration of molecules, which are detectable
due to the emissions from their radioactive labels, is added to the biological system
as a radiotracer to examine the metabolism, biokinetics, and biodistribution of these
molecules. The concentration of the radiolabelled tracer added to the biological system is small enough not to alter the properties of the process under investigation.
This presumes that the radiolabelling of a molecule does not alter its physiological
and biochemical properties since the radiotracer must behave in the system exactly
as does the native nonlabelled molecule. The only additional property of the tracer is
that it simultaneously broadcasts information, which allows a mapping of the physiological function and metabolism of a tissue and from this mapping conclusions on
tissue function or dysfunction can be drawn [1–4].

The widespread utilisation and growing demand for these techniques can be
attributed to the development and availability of a rapidly increasing number of
specific radiopharmaceuticals (currently more than 100) [3] and the availability
of suitable radionuclides produced in research reactors [5, 6] or with accelerators
[7, 8]. Whereas the number of research reactors is in slight decrease and facilities

U. Holzwarth
European Commission, Joint Research Centre, Institute for Health and Consumer Protection
T.P. 500, Via E. Fermi 2749, 21027 Ispra (VA), Italy
e-mail: uwe.holzwarth@jrc.ec.europa.eu

M.C. Cantone and C. Hoeschen (eds.), *Radiation Physics for Nuclear Medicine*,
DOI 10.1007/978-3-642-11327-7_6, © Springer-Verlag Berlin Heidelberg 2011

are increasingly ageing, with booming positron emission tomography (PET)[1] there has been a surge in the technical development and production of small-size, low-cost cyclotrons (with proton energies below 20 MeV) for the local production of short-lived PET radionuclides like ^{11}C, ^{13}N, ^{15}O, and ^{18}F [3,7]. Most of the more than 350 cyclotrons worldwide are involved in the production of ^{18}F for PET tracers like [^{18}F]-FDG (2-[^{18}F]-2-desoxy-D-glucose). ^{18}F can substitute an OH group in a biomolecule without significantly changing its biochemical physiological properties. With a physical half-life of about 110 min, ^{18}F-labelled molecules can even be distributed from a central radiopharmacy equipped with a cyclotron to satellite PET centres that can be reached within typically one half-life. Thus, today about 10% of all imaging procedures in nuclear medicine are performed with ^{18}F-labelled molecules [3].

For scintigraphy and single-photon emission computed tomography (SPECT), ^{99m}Tc is the most used radionuclide since it can be eluted at the hospital from a ^{99}Mo/^{99m}Tc generator [10].

The convenience of having a radionuclide generator that can be used for 1 week promoted the development of ^{99m}Tc radiochemistry and of *cold* kits, so called because they do not contain radioactivity. Cold kits are an efficient way to formulate ^{99m}Tc-labelled radiopharmaceuticals from sodium pertechnetate solution eluted from a ^{99m}Tc-generator. Cold kits are prepared on large scale by companies or laboratories in such a manner that they have long shelf lives of at least several months. Usually, they can be transported at room temperature and are then stored under refrigeration to ensure stability [3]. Dozens of different labelling kits are available that allow using ^{99m}Tc for a wide variety of diagnostic applications. Therefore, ^{99m}Tc is used in about 80% of the about 30 million imaging procedures per year worldwide [11].

Radioiodine (^{131}I) was first used in the treatment of metastasized thyroid cancer in 1943 [12] and remained the most efficient treatment of hyperthyroidism and thyroid cancer until today [3]; further optimizations in terms of increased response rate [13] and reduced risk of secondary primary malignancies [14] can still be expected. Also, the palliative treatment of pain from skeletal metastasis with

---

[1] PET radiopharmaceuticals use as labels radionuclides that undergo a *positron* (β^+)-decay. After the decay event the positron looses its kinetic energy by collisions with atoms and molecules. Together with an electron the now *thermalized* positron forms positronium, i.e., an unstable hydrogen-like state. After at maximum a few μs, electron and positron become annihilated and the annihilation converts the masses of both particles into two γ-quanta with an energy of 511 keV each. Due to conservation of both energy and momentum these γ quanta are emitted at an angle of about 180°. As these γ-rays have sufficiently high energy they penetrate the body with low probability of interaction and can be imaged externally by means of coincidence measurements in two detectors (of a ring of detectors around the patient) located on the opposite site of the patient. This coincidence measurement gives PET an advantage over SPECT because collimators are no longer required, yielding a 10 to 100 times higher sensitivity than SPECT, which is moreover independent of the desired resolution that depends mainly on the size of the individual detector elements. Since positron emitting PET radionuclides like C, ^{13}N, ^{15}O and ^{18}F do not significantly alter the properties of the tracer molecules their metabolism remains essentially unchanged and a quantification of their biodistribution requires only an input function that can be obtained from a blood sample [4,9].

radiopharmaceuticals such as ^{89}SrCl, ^{153}Sm-EDTMP (ethylenediamine tetra(methylene phosphonic acid)), ^{186}Re-HEDP (hydroxyethylidene diphosphonate), or ^{188}Re-HEDP is practiced clinically, and response rates of up to 80% are reported [15]. Advances in tumour biology and radiobiology, in the technology of antibody production and engineering [16], in peptide synthesis, and in radiochemistry [17] have enabled vast research on the development of targeted cancer therapies with radiopharmaceuticals [18]. Such radiopharmaceuticals should associate exclusively with tumour cells regardless of their location and kill or inactivate them. At the same time, the dose to healthy tissue should be kept to acceptably low levels [19]. This research has brought about the first two (approved by the Food and Drug Administration [FDA]), commercially available radiolabelled, antibody-guided pharmaceuticals (Zevalin^{TM} and Bexxar^{TM}) for the treatment of B-cell lymphoma [20–23]; these are supplied as a kit to be labelled with ^{90}Y or as a final product labelled with ^{131}I, respectively.

After a lot of effort has been spent on the in vitro quantification of the expression of somatostatine receptors in human neuroendocrine tumour tissues [24, 25], noticeable progress is made in the development of peptide receptor radionuclide therapies (PRRTs) [26–28]. Not only ^{18}F-FDOPA (fluorodihydroxyphenylalanine) has been proposed for diagnosis and staging [29], but also diagnostic radiopharmaceuticals are available to verify the receptor density in vivo for dosimetry and treatment planning. These are mainly labelled with ^{111}In [30]; recently, ^{68}Ga [30, 31] was introduced for this purpose, whereas ^{90}Y and ^{177}Lu are used in therapy (e.g., [2, 32]).

To improve the efficacy of antibody-guided radionuclide therapy, *pretargeting* strategies are being developed [2, 33, 34] that use either bispecific antibodies [35] or the avid interaction between avidin/streptavidin and biotin [36]. Bispecific antibodies bind with high affinity on one side to the tumour-associated antigens and on the other side to a radiolabelled hapten, which is administered in a later phase when the superfluous antibody construct has cleared from the bloodstream. The avidin/biotin approach usually applies a clearing agent in an intermediate step to remove the excess pretargeting substance from the bloodstream. The radiolabelled carrier is administered in a next step and binds with high affinity to the pretargeting substance. Generally, the radiolabelled molecules can be much smaller than antibodies and can therefore reach their targets faster. This allows a labelling with a shorter-lived radionuclide that provides a higher dose rate in the target tissue. Bivalent haptens, binding to more than one target molecule, can improve the efficacy of tumour targeting. Pretargeting molecules that provide more than one binding site for the hapten can multiply the receptor density for the radiolabelled conjugate. In this way, higher doses and dose rates can be achieved in the target tissues while increasing the dose ratio of tumour to normal tissue (e.g. [2, 37]).

To increase the amount of radioactivity on the treatment site, the emerging field of nanomedicine uses microspheres such as liposomes or polymeric micelles loaded with radioactivity or radioactive nanoparticles [38–40]. One promising application is the so-called radioembolic therapy of tumours in which the carriers are injected in the tumour vascularisation [41, 42]. In case of ferromagnetic nanoparticles, the radiation effect may be enhanced by combining magnetic targeting in a strong magnetic field gradient with mild local heating by applying an external alternating magnetic field to increase the radiosensitivity of the targeted tissue (cf. [43, 44]).

In spite of all recent and future developments, the challenge in radiopharmaceutical development and production will remain finding the proper combination of a carrier molecule or carrier system with a radionuclide that meets all medical and pharmacological requirements [1, 2, 16, 17]. The complete task is a highly multidisciplinary project and requires knowledge in different fields such as radiation physics, radiochemistry, biochemistry, biotechnology, immunology, oncology, pathology, haematology, radiobiology and dosimetry, pharmacology, and nuclear medicine [45, 46]. Especially, the radiochemical labelling procedure needs to take into account a large variety of biological and pharmacological factors [45]. The final result will be a compromise among the radiobiological properties of the radionuclide, the physiological and biochemical properties of the carrier or tracer molecule, and the radiochemical and chemical feasibility to produce a pure, compound that is stable in vivo and still exhibits the envisaged pharmacological properties and medical effect.

6.2 Radionuclide Properties

More than 1,000 radionuclides can be produced in nuclear reactors or with accelerators, but only a few of them offer physical and chemical properties that make them suitable for medical applications. But, the radionuclide with the best-suited physical properties cannot always be used for a specific tracer. Special attention has to be paid to the chemical properties that substantially influence the selection of the labelling method, which then decides on success or failure of the labelled compound.

6.2.1 Quality and Energy of the Emission

Nuclear medical imaging is based on the detection of emissions from a radionuclide that has been infiltrated in the metabolism of the body. Hence, one is seeking γ-ray emitters with a γ-ray energy of around 150 keV, for which the sensitivity and resolution of currently used SPECT cameras has been optimised[2] [47]. For PET, the energy of the emitted positrons should be as small as possible to minimise the distance between the emission and the annihilation sites of the positron. The travel distance of the emitted positron contributes to the physical resolution limit of PET imaging because it is the distance between the location of the radiotracer that is supposed to be imaged and the origin of the detected 511-keV γ-quanta that are

---

[2] This optimization is a consequence of the properties of ^{99m}Tc as the dominating radiolabel for non-PET diagnostics.

Table 6.1 Compilation of the most important PET radionuclides

Nuclide	Half-life	Average β^+-energy (MeV)	Mean range in tissue (mm)	β^+ intensity (%)
^{11}C	20.385 min	0.386	0.3	99.75
^{13}N	9.965 min	0.492	1.4	99.80
^{15}O	2.037 min	0.735	1.5	99.90
^{18}F	109.77 min	0.250	0.2	96.73
^{62}Cu	9.673 min	1.314	2.3	97.43
^{64}Cu	12.701 h	0.278	0.2	17.60
^{68}Ga	67.71 min	0.830	1.9	89.14
^{82}Rb	76.38 s	1.479	2.6	95.43
^{86}Y	14.74 h	0.660	0.7	31.9
^{89}Zr	78.41 h	0.396	0.3	22.74
^{124}I	4.176 days	0.820	0.8	22.7

All nuclear data were taken from [49]. The average β^+-energy is given because it is more significant than the frequently cited maximum energy of the β^+-spectrum for which the emission probability tends to zero. The data for the mean range in tissue have been taken from [4, 50].

Table 6.2 Compilation of the most frequently used radionuclides for planar gamma camera and single-photon emission computed tomography (SPECT) imaging and their most relevant physical properties

Nuclide	Half-life	Preferentially imaged γ-energy (keV)	Intensity (%)	Decay mode	Source
^{67}Ca	78.28 h	93.3	38.81	EC	Cyclotron
		184.6	21.41		
		300.2	16.64		
^{81m}Kr	13.10 s	190.5	64.9	IT	Generator
^{99m}Tc	6.015 h	140.5	89.06	IT	Generator
^{111}In	67.31 h	171.3	90.7	EC	Cyclotron
		245.4	94.1		
^{123}I	13.22 h	159	83.3	EC	Cyclotron
^{131}I	8.025 days	364.5	81.5	β^-	Reactor
^{133}Xe	5.243 days	81.0	38.0	β^-	Reactor
^{201}Tl	73.01 h	167.4	10.0	EC	Cyclotron

All nuclear data were taken from [49]. Only the most frequently imaged γ-rays are presented with their energy and intensity. The decay modes EC, IT, IC indicate electron capture, isomeric transition, and internal conversion, respectively.

imaged [4]; therefore, high positron energies will degrade spatial resolution [48]. The physical properties of the most important diagnostic radionuclides are compiled in Tables 6.1 and 6.2.

Radionuclides emitting β^-- and α-radiation are avoided in diagnostic applications because they cause excessive radiation exposure without any benefit for imaging. An exception to this rule are the so-called breath tests, which are performed using either the stable carbon isotopes ^{13}C (and less frequently ^{11}C) or the β^--emitter ^{14}C. The common principle of these tests consists of the oral administration of a substance labelled with one of the mentioned carbon isotopes. The substance is then metabolised, and the stable or radioactive carbon labels are incorporated in CO_2 molecules that can be found in exhaled breath. Their concentration is determined using a gas chromatograph isotope ratio mass spectrometer or scintillation detection, respectively. This method allows a precise evaluation of the presence or absence of etiologically significant changes in metabolism due to the lack of a specific enzyme or due to a specific disease [51], such as the presence of bacteria in the large or small bowel that metabolise the administered substance, thereby producing ^{14}C-CO_2, which is then taken up by the blood and exhaled. Among the various applications, mainly three tests are performed quite frequently: fat absorption using ^{14}C-trioleine, bile acid absorption using ^{14}C-chologlycine, and most commonly, the detection of *Helicobacter pylori* infection in the stomach using ^{14}C-urea [52]. Breath tests usually require collection of multiple breath samples over an extended time period, which creates difficulties in routine clinical practice (see the editorial [51] and articles in this special issue).

On the other side, β- and α-radiations are suitable for therapeutic applications in which a high radiation dose needs to be delivered to a limited target volume. A coemission of low-energy γ-rays or positrons at low intensity for imaging might be desired to assess the biodistribution of the radiopharmaceutical and the corresponding dose distribution, whereas emissions of high-energy γ-rays are undesired.

The emission of α-particles with a typical range of a few cell diameters would be ideal for the treatment of single tumour cells, residual disease, or micrometastasis [46,53], whereas the emission of β-particles (whose range varies, depending on their energy, typically between half a millimetre and a few millimetres) would be more suitable for larger metastasis and small solid tumours. By the proper choice of the β-energy, the absorbed dose delivered to the tumour can be optimised with respect to tumour size [53–55]. Also, Auger emitters can be used for therapy provided that the carrier molecules are internalised by the targeted cells and the radionuclide can get sufficiently close to their DNA, inducing enough damage to inactivate them. The biological effect of radiation is directly related to the linear energy transfer (LET), which is the average energy deposited by a particle per unit track length (keV/μm). High-LET radiation like α-particle radiation (25–230 keV/μm [56]) can destroy cells even by single hits, whereas low β-radiation (LET $\approx$ 1 keV/μm) is much less efficient. Auger emitters may reach the LET values of α-particles but confined in a range of about 10 nm, which explains why they should be targeted directly to DNA [19,57]. β-Particle radiation, however, has an advantage in bulky tumours with poor tumour penetration when the therapeutic compound remains mainly located on the surface of the tumour. In such cases, also cells of the inner tumour area may be killed by multiple hits from various directions due to β-*cross fire* emitted from the compound on the tumour surface (e.g. [12]). To increase the therapeutic impact, also the application of radionuclide *cocktails* is considered to better match dose distribution with tumour size [55,58].

Tables 6.3 and 6.4 compile the key properties of the most applied and frequently discussed therapeutic radionuclides. Both tables are not comprehensive.

Table 6.3 Compilation of β- and α-particle emitting radionuclides for radionuclide therapy

Nuclide	Half-life	Average β- and α-particle properties			γ-Rays used for imaging	Source
		Energy (MeV)	Range	Intensity	Energy (intensity)	
^{76}As	26.24 h	1.070	5.0 mm	100%	559 keV (45.0%)	Reactor
^{90}Y	64.00 h	0.934	3.9 mm	100%	None	Generator
^{188}Re	17.00 h	0.763	3.5 mm	100%	155 keV (15.6%)	Generator
^{166}Ho	26.82 h	0.665	3.2 mm	100%	80.5 keV (6.7%)	Reactor
^{32}P	14.26 d	0.695	2.9 mm	100%	None	Reactor
^{89}Sr	50.53 d	0.585	2.5 mm	100%	None	Reactor
^{186}Re	89.24 h	0.345	1.8 mm	100%	137.1 keV (9.5%)	Cyclotron
^{198}Au	64.68 h	0.312	1.6 mm	100%	411.8 keV (95.6%)	Reactor
^{77}As	38.83 h	0.226	1.2 mm	100%	None	Reactor
^{153}Sm	46.50 h	0.224	1.2 mm	100%	103.2 keV (29.3%)	Reactor
^{131}I	8.02 days	0.181	0.9 mm	81%	364 keV (82%)	Reactor
^{161}Tb	6.90 days	0.154	0.8 mm	101%	74.6 keV (10.2%)	Reactor
^{67}Cu	61.83 h	0.141	0.7 mm	100%	184.6 keV (48.7%)	Cyclotron
^{177}Lu	6.647 days	0.134	0.7 mm	100%	208.4 keV (10.4%)	Reactor
^{169}Er	9.39 days	0.100	0.5 mm	100%	None	Reactor
^{199}Au	75.34 h	0.082	0.4 mm	100%	158.4 keV (40%)	Reactor
^{213}Bi	45.6 min	8.320	85 μm	100%	440.5 keV (25.9%)	Generator
^{212}Bi	60.6 min	7.738	82 μm	100%	None	Generator
^{211}At	7.21 h	6.746	65 μm	100%	None	Cyclotron
^{149}Tb	4.12 h	3.967	28 μm	16.7%	165 keV (26.4%)	Cyclotron
^{226}Th	30.6 min	6.917	70 μm	400% (36.2 s)	None	Generator
^{225}Ac	10.0 days	6.867	69 μm	400% (50.4 min)	218.1 keV (11.4%/221Fr) 440.5 keV (25.9%/213Bi)	^{229}Th or Cyclotron
^{224}Ra	87.84 h	6.566	64 μm	400% (61.5 min)	241 keV (4.1%) 238.6 keV (43.6%/212Pb)	^{227}Th
^{223}Ra	11.435 days	5.668	53 μm	400% (36.2 min)	269.5 keV (13.9%) 351 keV (12.9%/211Bi)	^{227}Th

The β-particle emitters are arranged in order of descending range in tissue (data from [1, 59]). The range of α-particles for the average energy given was interpolated from the data given in [59, 60]. All other nuclear data were taken from [49]. Intensities for β- and α-emissions are normalised to the decay of the specified radionuclide. Selection criteria for γ-emissions suitable for imaging were (a) maximum energy not much higher than of ^{131}I (364 keV) and (b) at least 5% intensity. The α-particle emitters have complex decay schemes with cascades and branchings. The given energy is the average over all α-particles emitted until a stable nuclide is reached. Intensity 400% means that the specified radionuclide provides a cascade of 4 α-particles. The time given in brackets is the sum of all half-lives following the first decay in the cascade and indicates in which time span the α-particles are emitted.

6.2.2   Physical Half-Life

The physical half-life $T_{1/2}$ of the chosen radionuclide has to match the biokinetics for a given application. In diagnostics, it should be just long enough to complete the

Table 6.4 Compilation of Auger and conversion electron emitters in use or in discussion for radionuclide therapy

Nuclide	Half-life	Average e⁻ properties		Decay mode	γ-Rays used for imaging		Source
		Energy	e⁻/decay		Energy	Abundance (%)	
^{51}Cr	27.70 days	3.97 keV	4.68	EC	320 keV	10	Reactor
^{67}Ga	78.28 h	7.07 keV	7.03	EC	SPECT		Cyclotron
^{77}Br	57.04 h	4.13 keV	4.96		239 keV	23.1	Cyclotron
^{94}Tc	4.883 h	5.17 keV	6.42	EC	PET	10.5	Cyclotron
^{99m}Tc	6.01 h	0.96 keV	4.67	IT	141 keV	89	Generator
^{111}In	2.82 days	6.51 keV	6.05	EC	SPECT		Cyclotron
^{114m}In	49.51 days	4.15 keV	7.74	EC	558 keV	3.2	Reactor
^{115m}In	4.49 h	2.85 keV	5.04	IT, β⁻	336 keV	45.8	Reactor
^{123}I	13.20 h	7.33 keV	12.6	EC	SPECT		Cyclotron
^{124}I	4.176 days	4.87 keV	8.6	EC, (β⁺)	PET		Cyclotron
^{125}I	59.40 days	11.9 keV	21.0	EC	Not suitable		Cyclotron
^{167}Tm	9.25 days	13.6 keV	11.4	EC	207.8 keV	42	Cyclotron
^{193m}Pt	4.33 days	10.9 keV	20.3	IT	None		Reactor
^{195m}Pt	4.01 days	21.8 keV	31.5	IT	98.9 keV	11.7	Reactor
^{201}Tl	73.01 h	15.27 keV	36.9	EC	SPECT		Cyclotron
^{203}Pb	51.92 h	11.63 keV	23.3	EC	279.2 keV	80.9	Cyclotron

Data for the abundance and average energy of the electrons per decay were retrieved from [55] and [19]. Other nuclear data and γ-emissions suitable for imaging and dosimetry have been collected from [49].

The decay modes EC and IT indicate electron capture and isomeric transition, respectively. *PET* positron emission tomography, *SPECT* single-photon emission computed tomography.

imaging procedure. For therapeutic applications, the $T_{1/2}$ should be ideally about two to three times longer than the time required for achieving maximum uptake of the radiopharmaceutical in the target tissue [61]. On the other hand, the $T_{1/2}$ must be short enough to ensure a sufficiently high dose rate in the target cells to avoid repair mechanisms becoming successful in targeted cells. Theoretical calculations considering effects of cell proliferation indicate that even much longer living radio-labels (e.g., ^{91}Y [58.51 days] instead of ^{90}Y [64.00 h]) could provide a therapeutic advantage [62, 63].

Incorporated radionuclides are also characterised by their biological half-life T_{bio}, which is given by the time required to excrete half of the substance. T_{bio} depends also on the chemical form in which the radionuclide is present and that affects the metabolic pathways of excretion. The combined effect is described by the effective half-life, defined as

$$T_{eff} = \frac{T_{bio} \cdot T_{1/2}}{T_{bio} + T_{1/2}} \tag{6.1}$$

In other words, if the travelling time of the carrier molecules in the body is too long, the molecules will reach their targets after the radionuclide has already decayed.

This will cause an unspecific, unacceptable radiation dose to healthy tissues. If the same carrier molecule would be labelled with a much longer-lived radionuclide, practically all carriers would deliver their radioactive charge to the target cells, but the dose rate might be too low for achieving a therapeutic effect.

From Table 6.3, it is evident that all α-emitters unfortunately have very short half-lives. Therefore, the concept of in vivo generators has been suggested, which means using long-lived α-emitters like ^{225}Ac ($T_{1/2} = 10$ days) or ^{227}Th ($T_{1/2} = 18.7$ h) that decay in a sequence of short-lived α-emitting daughter isotopes [64]. The approach depends on the possibility to control the localization of all daughters, step by step, since it is questionable whether the label will stay attached to its carrier due to recoil effects and possible loss from the chelator. The shorter the time span in which all α-particles are emitted, the higher is the probability of achieving localization control. From the data indicated in Table 6.3 (cf. [65–67]), it appears that this goal might only be achievable for ^{226}Th.

6.2.3 Specific Activity and Purity

Radionuclides for pharmaceutical applications must be available with high purity, high activity concentration, and a high specific activity. The specific activity of the radionuclide concerned $A_{s,i}$ is defined as

$$A_{s,i} = \frac{A_i}{m_i + \sum_{j \neq i} m_j} \tag{6.2}$$

that is, as its activity A_i divided by its mass m_i plus the sum of the masses of all other radioactive or stable isotopes of this element that are present in a chemical form that makes them competitors in the synthesis process of the radiopharmaceutical [68]. Molecules with improper labels will competitively bind to the receptor site and may obscure the binding of the radiolabelled ligand, thereby compromising its therapeutic efficacy, since the capacity of the targeted receptors on malignant cells may be as low as a few nanomoles. Thus, specific activities ≥ 70–200 GBq/μg are required [2].

The theoretical maximum specific activity can be calculated assuming that one mole of the isotope consists completely of the desired radioactive species as

$$A_{s,\max,i} = \frac{\lambda_i N_A}{m_{\mathrm{mol},i}} \tag{6.3}$$

where λ_i is the decay constant of the radionuclide, $m_{\mathrm{mol},i}$ denotes its molar mass, and N_A is the Avogadro number. This theoretical value is difficult and sometimes even impossible to achieve for physical reasons [68, 69].

The specific activity is often difficult to determine since also the content of nonradioactive isotopes needs to be quantified either by methods of mass spectrometry or by neutron activation analysis [69]. The coproduction of other stable or radioactive isotopes of the same element that cannot be separated by chemical methods limits the obtainable specific activity. Therefore, the chemical and isotopic composition of the target and the choice of the production route require a lot of attention [7].

Also, the terms *carrier free* and *no carrier added* are frequently used. *Carrier free* means that the preparation of the radionuclide is free of other stable or radioactive isotopes of the element and should theoretically exhibit the maximum specific activity. *No carrier added* is a more cautious term that indicates that during the preparation attention was paid to avoid any contamination with the element in question in the same chemical form enabling isotopic exchange reactions [68].

6.2.4 Availability of Radionuclides

Radionuclides for routine clinical use should be readily available and inexpensive. Neutron-rich, β^--emitting radionuclides are mainly produced in reactors by neutron capture reactions or as fission products and are usually cheaper than accelerator-produced radionuclides. However, this production route is often related to a limited specific activity since parent or target isotopes and the (n, γ)-reaction product are chemically identical and cannot be separated [45, 69]. The probability of (n,p)- or (n, α)-reactions that result in chemically different reaction products is much lower; hence, such radionuclides are more expensive, especially if isotopically enriched target materials are required, which is the normal situation for accelerator-produced radionuclides [45, 70].

Radionuclide generators are a convenient way for producing no-carrier-added β- and α-emitting radionuclides [10]. They employ a relatively long-lived parent radionuclide that decays into a more short-lived daughter radionuclide. Due to the different chemical properties of parent and daughter nuclide, the daughter can be chemically separated [10, 45]. Most commonly, the parent nuclide is bound to a chromatographic ion exchange column, and the daughter is eluted using a liquid in which it is soluble [10]. Table 6.5 compiles the properties of some of the more important radionuclide generators.

The most used generator system is the ^{99}Mo/^{99m}Tc generator, in which molybdate $\left(^{99}\text{MoO}_4{}^{2-}\right)$ is bound to an aluminium oxide column, which is then shipped in a shielded container to hospitals, where it can be used for up to 6 days [71]. ^{99m}Tc is extracted by elution with physiological saline solution. The radiopharmaceutical is frequently prepared using cold kits. ^{99m}Tc decays by γ-ray emission into ^{99}Tc, which is a pure β^--emitter with a half-life of 2.1×10^5 years and does not much contribute to radiation exposure [71]. ^{99}Mo is produced as a fission product from highly enriched (up to 97%) ^{235}U targets with a fission yield of 6.11%. Typical irradiation conditions are up to 6 days with a thermal neutron flux of $10^{14}\,\text{cm}^{-2}\,\text{s}^{-1}$ [71]. After 6 days of cooling, the targets are shipped for radiochemical processing. The

Table 6.5 Compilation of important radionuclide generator systems and their properties

Generator system	Half-life		Daughter isotope		Generator production
	Parent isotope	Daughter isotope	Application	Decay mode	
^{99}Mo/^{99m}Tc	67 h	6.01 h	SPECT	IT	Reactor
^{81}Rb/^{81m}Kr	4.58 h	13.3 s	SPECT	γ	Cyclotron
^{68}Ge/^{68}Ga	270.8 days	67.7 min	PET	β^+	Cyclotron
^{82}Sr/^{82}Rb	25 days	76.4 s	PET	β^+	Cyclotron
^{90}Sr/^{90}Y	28.5 year	64.0 h	Therapy	β^-	Reactor
^{188}W/^{188}Re	69.4 days	17.0 h	Therapy	β^-	Reactor
^{224}Ra/^{212}Bi	3.66 days	60.6 min	Therapy	β^-	^{228}Th
^{225}Ac/^{213}Bi	10.0 days	45.6 min	Therapy	α	Cyclotron, reactor
^{230}U/^{226}Th	20.8 days	30.6 min	Therapy	α	Cyclotron

Nuclear data from [49]. *PET* positron emission tomography, *SPECT* single-photon emission computed tomography.

radiochemical separation of the dissolved uranium targets yields carrier-free ^{99}Mo of very high specific activity, which allows production of small-volume columns and elution of ^{99m}Tc in a small volume [10].

Short-lived PET radionuclides such as ^{11}C, ^{13}N, and ^{15}O require a cyclotron on site. The only PET radionuclide that allows a pseudoisotopic labelling of biomolecules, by replacing an OH$^-$ group, and that can be supplied within certain limits from a central radiopharmacy is ^{18}F. In addition, more long-lived ones like ^{89}Zr and ^{124}I for antibody labelling are becoming more popular. Recent improvements of the performance of ^{68}Ge/^{68}Ga generators now enable the labelling of peptides with specific activities of more than 50 MBq/nmol [72]. This has been achieved by reducing the contents of metallic impurities [Zn(II), Ti(IV), Fe(III)] and gallium in the wrong oxidation state [Ga(IV)] by orders of magnitude and by reducing the eluate volume [72].

6.3 The Physics of Radionuclide Production

Radionuclides are produced by neutron reactions in a nuclear reactor, providing neutron-rich radionuclides, or by interaction with charged particles, usually accelerated with cyclotrons, which provides neutron-deficient isotopes. Important medical radionuclides like ^{99}Mo and ^{131}I are radiochemically separated from neutron-irradiated nuclear fuel. For more details on reactor-produced and accelerator-produced radionuclides, refer to the *TECDOC 1340 – Manual for Reactor Produced Radioisotopes* [5] and *Technical Report Series No 465 – Cyclotron Produced Radionuclides: Principles and Practice* [7], respectively. Both documents were published by the International Atomic Energy Agency (IAEA) and present an excellent overview of physics, technology, and nuclear data.

6.4 The Technology of Radionuclide Production

6.4.1 Reactor Versus Accelerator Production

The production of ^{99}Mo in nuclear reactors for ^{99m}Tc generators is described in the section "Availability of Radionuclides." It has been stated that ^{99}Tc, like the other important fission product ^{131}I, can be well separated chemically from the dissolved uranium target, yielding a clean product with high specific activity. For the alternative ^{99}Mo production by ^{98}Mo$(n, \gamma)^{99}$Mo, this advantage is lost [10]. Neutron capture reactions have the disadvantage that target and product are chemically identical and cannot be separated, which limits the achievable specific activity. Only in rare cases is it possible to produce a radionuclide by neutron capture that subsequently decays rapidly in the desired radionuclide, as in the case of ^{177}Lu, which can be produced via the reaction ^{176}Lu$(n, \gamma)^{177}$Lu on isotopically enriched ^{176}Lu targets [73] or by the process ^{176}Yb$(n, \gamma)^{177}$Yb $\rightarrow \beta^- \rightarrow\ ^{176}$Lu [74, 75] as a way to start with a chemically different target material.

Nuclear reactions with charged particles, such as protons, deuterons, α-particles, or ^{3}He^{2+} ions usually provide a product that is a different element from the target material and facilitates preparations with high specific activities. Moreover, the coproduction of radionuclidic impurities can be controlled to a certain extend by the proper choice of the particle energy window.

Nevertheless, the basic consideration that neutron-rich radionuclides are usually obtained form nuclear reactors whereas neutron-deficient isotopes are produced by accelerators shows that both sources complement each other rather than being in competition for providing the whole range of medically interesting radionuclides [7].

6.4.2 Cyclotrons

Light ions are supplied by a plasma ion source placed in the centre of the cyclotron (see Fig. 6.1). The ions are then accelerated along a spiral trajectory that is guided in the easiest case inside two half-cylindrical hollow electrodes traditionally referred to as *dees* due to their original shape. They are placed in a vacuum chamber, which is itself placed between the pole pieces of a powerful electromagnet that creates a static magnetic field perpendicular to the particle orbits. Inside the dees, the particles are only affected by the magnetic field; the electrical field that accelerates them is only effective in the moment the particles are passing the gap between the dees. In a constant magnetic field, the radius of the particle orbits increases with energy, and the higher velocity of the particles just compensates for the longer distance per turn in a way to complete every turn always in the same time, independent of the current particle energy. This is referred to as *isochronous* orbits. In this way, the particles can be accelerated by applying an alternating high-voltage field with

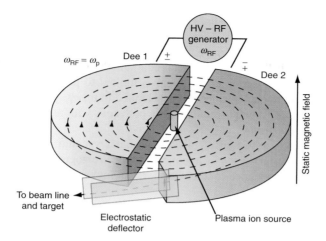

Fig. 6.1 Schematic presentation of the working principle of a cyclotron. In the centre between the dees, a plasma ion source supplies the ions to be accelerated. In the gap between the dees, an alternating high-voltage field is applied that accelerates the particles on every passage. A magnetic field perpendicular to the orbit plane forces the particles on a circular orbit whose radius increases after each acceleration step. An electrostatic extraction system helps to extract the particle beam from the cyclotron. *HV-RF:* high-voltage radio frequency

constant frequency, and they will always see the proper polarity and magnitude of the electrical field when passing the gap between the dees. Typical frequencies are in the range between 10 and 30 MHz. This principle is working for classical particles, that is, those whose total energy is increasing by an increase of kinetic energy and not yet significantly affected by a relativistic mass increase. The maximum particle energies required for most pharmaceutical radionuclide productions are sufficiently low ($\leq$20 MeV) not to be affected by significant relativistic effects. Some means of focusing is requested; otherwise, particles that start from the ion source under a small angle with the orbit plane will spread out into the dees and get lost. Initially, at low particle energies, the focussing is supplied by the accelerating electric field, whereas after the particles have gained energy the focusing is achievable by a slight weakening of the magnetic field towards the peripheries of the dees. However, particles that experience a slight relativistic mass increase rather require a stronger magnetic field. Thus, another focussing method is applied that uses an azimuthally varying magnetic field by milling spiral segments in the iron of the electromagnet that result in segments of a lower magnetic field where the distance between the iron is larger (valleys), and of a higher field where the iron is closer (hills). In this way, the magnetic field can even be increased slightly towards the periphery to compensate for relativistic effects, and energies up to 500 MeV can be realised with such cyclotrons [7].

To extract the particles from the cyclotron, on their last turn they pass an electrostatic *deflector*, which is a curved capacitor in which a high-voltage electrical field is applied that partially compensates for the effect of the magnetic field and deflects them towards the periphery, where they enter into the beam line. More modern cyclotrons accelerate particles as negative ions because it simplifies the extraction technique. The electrons are stripped when the ions pass a thin mica foil. As a result of the altered charge, the particles do not continue their trajectory in the magnetic field and leave the cyclotron. This system is limited to particles that form stable negative ions; helium nuclei cannot be accelerated in this way. Moreover, negative ion cyclotrons require a better vacuum since collisions with residual gas molecules may strip off the electrons.

The beam line is a system of tubes kept under vacuum; the beam is guided along the beam line and shaped by a sequence of electromagnetic quadrupole lenses, steering and bending magnets, and collimators until it reaches the target.

6.4.3 Targetry

The target is the place where the particle beam interacts with matter, thereby producing radionuclides. Whenever possible, the target is kept isolated from the vacuum of the beam line and the cyclotron by thin metal foils called *windows*. This is mandatory for liquid and gaseous targets and desired for solid targets, especially when the reaction products are volatile and could contaminate the beam line and cyclotron. This separation is problematic for irradiations using helium nuclei due to their high stopping power and the resulting energy loss in the window material.

Since the whole beam energy is dumped in the target, excellent cooling is required. For example, a proton beam with an energy of 20 MeV and an intensity of 50 μA has a power of 1 kW, which is usually distributed over a surface area of only 1 cm in diameter. In liquid targets such as those for the production of ^{18}F by proton irradiation of $H_2{}^{18}O$, the windows have to resist high pressure. To keep the energy loss in the windows low, they have to be thin and need a high mechanical strength at elevated temperatures. Windows under such working conditions are normally cooled by a helium jet from the beam side, which requires a second window to keep the helium separated from the beam line vacuum.

The desired properties of solid targets are a high melting point and a high thermal conductivity of the target material and a good thermal contact to a possible substrate material. Water cooling from the rear side is indispensable, and a helium jet cooling from the front side is required, especially if the reaction products are volatile. Moreover, a target material should be easy to process for isotope recovery, which is facilitated by a small volume. To increase the production yield, to avoid the coproduction of impurities, and to achieve high specific activity, pure and isotopically enriched target materials should be preferred. In many cases, compromises are required, for example, when oxide powders have to be used as target materials that often exhibit high melting temperatures but poor thermal conductivity. Metallurgical aspects may play a role if the product forms intermetallic phases with the target material that may affect later target processing. Generally, the target temperature should be kept reasonably low. In critical cases, a larger target surface can be irradiated with a diverging beam or the target being inclined under a small angle to the beam to distribute the beam energy on a larger surface and to facilitate cooling.

A homogeneous distribution of the beam intensity over the surface to be irradiated requires controlling the centricity of the beam on the target and of the beam intensity profile, which should be of Gaussian type. Care has to be taken that the ion beam does not unintentionally touch parts of the target system that are not supposed to be exposed. Apart from unwanted activation of the system, material can be transferred from the target system to the material to be irradiated that can compromise subsequent target processing and even radiolabelling.

Routine radionuclide production requires a type of remote handling of the target at least until it is placed in a shielded transport container. Liquid and gaseous targets offer more comfort than solid-state targets since the target content can be transferred to hot processing cells through properly shielded tubing systems. The material in contact with the radioactive load needs to be selected carefully to avoid contamination or retention.

An excellent compilation of target systems, the physics, engineering, and chemistry of targets is given in [7].

6.4.4 Radionuclide Isolation and Purification

The isolation and purification of the radionuclide from the irradiated target material is considered the first manufacturing step of a radiopharmaceutical. The methods applied vary with the target type and make use of different physical or chemical properties of the mother and daughter element. Since high amounts of radioactivity are handled corresponding to only tiny amounts of substance, precipitation techniques cannot be applied since they would require adding a large amount of an isotopic carrier, yielding a low specific activity of the desired radionuclide [76].

Distillation techniques can be applied if the vapour pressure of the target matrix and the produced radionuclide are sufficiently different at a given temperature. A special mode is thermal diffusion or dry distillation [7]. This method allows, for example, the extraction of ^{211}At from a Bi target produced by the reaction ^{209}Bi$(\alpha, 2n)^{211}$At (e.g., [77, 78]) or of ^{72}Se from a Ge target irradiated with α-particles [79].

If the parent material can be prepared as a liquid and the daughter is a gas, it can be separated by bubbling an insoluble carrier through the solution. An example in clinical use is the ^{81}Rb/^{81m}Kr generator [10].

A solvent extraction can be performed if parent and daughter are soluble in two different solvents that are immiscible, like water and an organic solvent [80]. This method can also be applied to extract ^{211}At produced via the reaction ^{209}Bi$(\alpha, 2n)^{211}$At from a dissolved Bi target and clean the product from coproduced Po impurities [81].

Today, ion exchange is the dominating separation technology because it is easy to handle and to shield. The desired radionuclide is absorbed on a chromatographic column. Such columns are mainly made from mineral column materials like Al_2O_3 that offer high resistance to radiation damage; also organic column materials are used that can be applied in fast processes when the column can be disposed as waste material following the processes (e.g. [82, 83]).

In some rare cases, the target material and the production route can be varied to obtain the desired radionuclide in a different chemical form. For example, the irradiation of ^{18}O-enriched water by the reaction $^{18}O(p, n)^{18}F$ yields ^{18}F in the chemical form of fluoride [84], whereas ^{18}F-F fluorine is obtained by the reaction $^{20}Ne(d, \alpha)^{18}F$ with neon gas to which traces of fluorine are added. If hydrogen gas is added, $H^{18}F$ is obtained [9].

6.5 Manufacturing of Radiopharmaceuticals

6.5.1 The Selection of Carrier and Tracer Molecules

In diagnostic and therapeutic applications, a tracer or carrier molecule is selected after having identified a molecular target that is uniquely related to the pathological dysfunction of the tissue to be imaged or eradicated. In this sense, FDG-PET uses a rather unspecific tracer since energy consumption per se is not a disease-related phenomenon. In spite of being successful in oncology, there is the desire for more tumour-specific tracers that behave more distinctly different in healthy and pathological tissue, such as tracers for hypoxic regions in cancer [85]. In therapy, the main targets are antigens [2, 33, 86] or peptides, such as somatostatin receptors [2, 28], gastrin-releasing peptide [28], or gastrin/cholecystokinin [28], that are uniquely expressed on malignant cells of certain cancers or at least overexpressed by orders of magnitude with respect to healthy tissue.

Once an adequate target and a carrier molecule that specifically recognises the molecular target structure and binds to it with high affinity are identified, the next step is to search for adequate radioactive labels that match the biokinetics of the carrier molecule. Biokinetics usually excludes labelling of antibodies with short-lived radionuclides for systemic applications (i.e., the radiopharmaceutical is injected intravenously) since the residence time of the antibody in the bloodstream may range from many hours to some days. There are two exceptions. The first one is target cells that also circulate in the bloodstream, as in haematological malignancies [86], or infected cells [87], such as in the case of HIV [88]. Another exception is locoregional administration, which means that the radiopharmaceutical is either directly injected in the tumour or in its vicinity or if residual disease is treated by injection in the cave created by surgical removal of the primary tumour (e.g., [1,89]). Hence, the administration mode has to be considered when developing a radiolabelled pharmaceutical. The choice of the radioactive label has of course to take into account all aspects mentioned in the section on radionuclide properties.

The next problem is to preserve the characteristics of the carrier molecule during radiolabelling.

6.5.2   General Labelling Requirements

Independently of a special labelling strategy chosen for a given case, the following aspects need to be considered (modified from [45]):

- The yield of the labelling procedure has to be maximised since the available activity of the radionuclide may be limited, and it contributes to a significant part to the overall price of the radiopharmaceutical.
- The procedure has to provide a radioconjugate of high purity. In general for a therapeutic application, the purity must be significantly higher than what may still be acceptable for a diagnostic conjugate.
- The specific activity should be as high as possible.
- The radiolabelled conjugate must exhibit sufficient in vivo stability and must not dissociate or be catabolized before the desired uptake in the target tissue has been achieved or before the imaging procedure has been completed.
- The labelling procedure must preserve the biochemical and physiological properties of the carrier molecule, especially its specificity and affinity to its target.
- The labelling method has to consider that linkers and chelators, as well as the radiolabel itself, may appreciably increase the molecular weight of the conjugate, which slows its biokinetics compared to the unlabelled carrier.
- In view of the high activities involved and the related dose rates, the labelling reaction and the subsequent purifying steps should be carried out under remote control in hot cells and if possible fully automatic under computer control.
- To facilitate its introduction into clinical practice, the labelling procedure should use as much as possible cheap and readily available materials and consumables. The number of processing steps should be reduced to a minimum and in that way minimise the risk of human error.
- In spite of striving for the highest specific activity and minimum volumes involved in processing, the risk of radiolysis has to be considered, especially in therapy when high-LET radionuclides are labelled to sensitive carrier molecules like peptides and antibodies.

6.5.3   Labelling Methods and Their Impact on Performance

Due to the exceedingly small quantities involved in radiochemistry, equilibrium reactions are not applicable for radiolabelling procedures [9]. One needs to keep in mind that, for example, the administration of a typical dose of ^{18}F-FDG for a whole-body scan of 370 MBq corresponds only to 1.1 ng of ^{18}F. Hence, reactions that have a complete turnover of the reactants are required.

Three basic types of chemical bonding are involved in radiolabelling [76]: (1) covalent bonding, in which each atom denotes one electron to the bond. Covalent bonds are very stable and dissociate in solution only as a result of a chemical reaction. Covalent bonding can be realised by isotopic labelling, for example, with the light PET radionuclides ^{11}C, ^{13}N, ^{15}O, and [18]F. The synthesis of ^{18}F-FDG makes use of a nucleophilic substitution reaction of an OH^--group by fluoride that exhibits high yields and good reproducibility. The majority of ligands used for non-PET applications cannot be labelled isotopically and are hence labelled nonisotopically by complexation of an isotope of a foreign element. In these cases, other bonding strategies are required. (2) Coordinate covalent bonds are formed when one atom donates both electrons to the bond, and (3) chelation occurs when more than one atom donates electrons to a labelling atom, often a transition metal.

For the production of iodine-labelled conjugates, one distinguishes direct and indirect radioiodination. Under proper control of the pH, the first method attaches iodine to the aromatic group of tyrosine in a fast reaction with high labelling yield. In case the molecule to be labelled does not contain tyrosine or the tyrosine is involved in the antigen recognition or the molecule exhibits vulnerable disulphide bonds, indirect labelling with intermediate linkers must be applied [45]. The indirect labelling gives lower yield but allows modification of properties such as intracellular retention or the excretion pathway of the catabolites by an intelligent choice of the linker molecule (for more details, see [45, 90, 91]).

A look at Tables 6.3 and 6.4 shows that most of the radionuclides used in therapy have a metallic nature and do usually not form covalent bonds with those elements that constitute the carrier molecule. Therefore, bifunctional chelators are used that form a noncovalent bond (chelate) with the metallic radiolabel and that can be bound covalently to the carrier or tracer molecule as schematically shown in Fig. 6.2. Since the chelation is a reversible process, only those pairs with the smallest dissociation constant

$$K_d = [M][L]/[ML] \qquad (6.4)$$

can be considered, where [M], [L], and [ML] denote the thermal equilibrium concentrations of the free metal, the free chelator, and the chelate–metal complex, respectively [45]. Moreover, the competition between chelating blood plasma

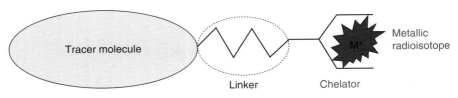

Fig. 6.2 Principle of labelling a tracer (diagnostics) or carrier (therapy) molecule, which may be an antibody, antibody fragment, a peptide, and so on with a radiometal incorporated in a bifunctional chelator that is attached by a linker

proteins that are present in much higher concentrations than the chelator in the conjugate ([L] + [ML]) requires chelators with K_d values that are orders of magnitude lower than those of their in vivo competitors.

The more stable the chelators are, the longer are the reaction times required for loading them with the radionuclide, which sometimes necessitates elevated temperatures that might be a risk for sensitive carrier molecules. In such cases, labelling with a preloaded chelator and subsequent purification might be a solution. Further attention must be paid to the effect the metal–chelate complex has on the kinetics and uptake of the radiolabelled molecule since large variations of mass and geometry may compromise the kinetics, the specificity, and the affinity of the molecule to the target. Therefore, the determination of the biokinetics and biodistribution of a new conjugate play an essential part in the radiopharmaceutical development process.

Figure 6.3 shows as an example for chelators of the DOTA (1,4,7,10-tetraazacyclododecane-1,4,7,10-tetraacetic acid) molecule family that is widely used for radiolabelling of short peptides [2, 27] since it offers high in vivo stability and some variability of the molecular structure by varying the amino acids in positions R1 and R2 as indicated in Table 6.6 [31]. In this example, it has been found that even different kinds of radionuclides attached to the peptide via the same chelator can affect the binding affinity to somatostatin receptors. Gallium radionuclides provide a significantly higher binding affinity to somatostatin receptor subtype 2

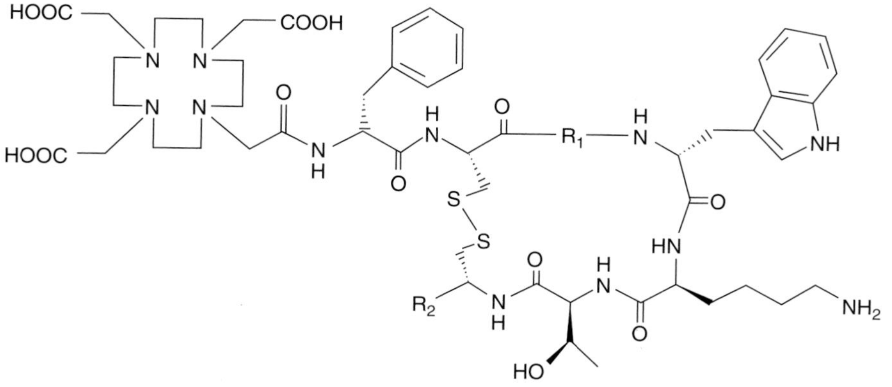

Fig. 6.3 Structural formula of the DOTA molecule family used for labelling of somatostatine analogues. (From [31])

Table 6.6 Some members of the DOTA family with the groups R1 and R2 in Fig. 6.3 (From [31]). Tyr and Thr denote the amino acids tyrosine and threonine, respectively

Compound		R1	R2
DOTA-TOC	[DOTA^0,Tyr^3]-octreotide	Tyr	Thr(ol)
DOTA-TATE	[DOTA^0,Tyr^3,Tyr^8]-octreotide	Tyr	Thr
DOTA-NOC	DOTA-1-naphtyl-alanine	Nal-1	Thr(ol)
DOTA-BOC	DOTA-benzothienyl-alanine	BzThi	Thr(ol)

(sst2) than the same molecular entity labelled with indium, yttrium, or lutetium radionuclides [31]. ^{67}Ga-DOTA peptides also showed a much higher internalisation into sst2-expressing cells [31]. This means that the effect of the labelling method on target affinity and internalisation needs to be carefully studied even for supposedly minor changes in synthesis and labelling because it will decide on the sensitivity of tumour detection in diagnostic applications and on the fraction of applied radioactivity that really reaches its target. The effect on internalisation of the compound into the cell moreover determines whether targeting with Auger emitters is feasible.

In the case of small radiolabelled molecules like peptides, the chelator and the radiolabel make up a large part of the molecular mass and influence the physicochemical properties of the compound, such as overall charge, lipophilicity, or hydrophilicity. For example, when the direct labelling of ^{131}I to octreotide for assessing the peptide receptor status was changed to an ^{111}In label coupled via DTPA (diethylene triamine pantaacetic acid), the compound became more hydrophil and the excretion pathway changed from hepatobiliary to renal, which reduced interfering radioactivity in the abdominal area and increased the diagnostic sensitivity of radiolabelled octreotide [45, 92]. This example shows that the polarity of a compound, and depending on it, its excretion pathway can be influenced by the labelling method, especially by the intelligent use of chelators and linkers. This provides an additional possibility to adjust the biodistribution of tumour targeting peptides [45].

Generally, the labelling method must avoid any chemical intervention (1) on molecular groups that are critical for the three-dimensional structure of the molecule and (2) of the amino acids that are critical for the molecular recognition and binding to the target. The first is especially critical for labelling of antibodies. For example, rhenium is a thiophilic element, and direct labelling of rhenium isotopes to antibodies will preferentially interact with the disulphide bridges that stabilise the molecule structure. As a result, the structure of the antibody might become distorted, and its binding properties will be degenerated [45]. In many cases, the success of labelling depends sensitively on the oxidation state of the radiometal and the reaction medium, and the reaction conditions need to be selected and stabilised carefully [45, 93].

Radionuclides for therapy should moreover possess good residualizing properties; this means that they should stay with their target until they decay. This can be achieved with nondegradable linkers or chelators that remain trapped inside the cell after the carrier molecule has been internalised and degraded. Trapping inside the cells requires radiocatabolites that are not soluble in phospholipids and hence cannot diffuse through the cell membrane, which allows only slow excretion by exocytosis [45].

A radio conjugate can lose its functionality also by radiolysis (e.g., [94]); that is, it is damaged by radiation during labelling or later during storage or transport. Protection against radiolysis is easier in the latter case, for which dilution or freezing of radiolabelled antibodies might be sufficient [95]. However, during labelling a small volume is required. It has been demonstrated that sensitive antibodies can successfully be protected even against damage by α-radiation by adding ascorbic acid (e.g., [96, 97]), which is an approved drug and does not interfere with chelation [45].

6.6 Quality Control of Radiopharmaceuticals

In contrast to conventional pharmaceuticals, radiopharmaceuticals have no or only a very short shelf life. This means that they cannot undergo complete quality control (QC) after production. The *Pharmacopoeia* [98] strongly recommends executing all tests that can provide results in a reasonable time. But, especially short-lived PET tracers have to be released before important QC tests, such as for sterility, can provide their results. In such cases, alternative procedures have to be defined and agreed on with the regulatory authorities to ensure that the product is always supplied sterile. This requirement can be met by adequate closed production and dispensing procedures in which the sterile pharmaceutical is produced from sterile ingredients transferred in presterilised containers, thereby preserving sterility by avoiding any contact with a nonsterile environment. Otherwise, a sterilisation procedure is necessary that requires additional time and equipment.

In general, the quality of radiopharmaceuticals relies on a meticulously elaborated quality assurance (QA) system rather than on postproduction QC. QA is based on well-defined and validated production and test procedures that can guarantee the purity and integrity of the dispensed manufactured product. If full QC is not possible before product release, quick and convincing test procedures have to be elaborated and validated that allow a decision regarding whether the radiopharmaceutical product complies with its specifications. If regulatory authorities agree, also a release "in bond" can be arranged that becomes effective when the user gets a final written release by fax. In such cases, the time available for QC can be extended by the transport time.

The radiopharmaceutical QC concerns the following parameters (according to [99]):

- The activity of the desired radionuclide has to be guaranteed within $\pm10\%$ for the specified batch or dose. This requires an absolute measurement of the activity with an accuracy of 2–5%. For this purpose, ionisation chambers are used that require a calibration with certified standard sources of adequate activity, energy, and geometry (shape and volume). The calibration must determine the calibration factor, the sample geometry factor, and the dynamic range accuracy.
- The nuclide identity can be determined by γ-ray spectrometry and the determination of its half-life. The γ-ray spectrometry also gives information on radioactive impurities. Such measurements are usually performed with high-purity Ge detectors that are properly calibrated for specified measurement geometry in γ-ray energy and detector efficiency.
- The radionuclidic purity is given as the ratio of the percentage of the radioactivity of the radionuclide concerned to the total radioactivity of the source. It is determined by γ-ray spectrometry. Standards are given in monographs on all commercial radiopharmaceuticals. The value depends on the production method and the production route of the radionuclide, as well as on the chemical purity of the target material. If no chemical cleaning or separation procedure can be performed, the radionuclidic purity can only be controlled using isotopically

enriched target materials and the proper choice of the nuclear reaction. If the half-lives of the impurities are longer than that of the radionuclide concerned, the impurities gradually enrich in the product and may limit the time within which the radiopharmaceutical can be administered.

- The radiochemical purity is the ratio expressed as a percentage of the radioactivity of the radionuclide concerned that is present in the chemical form declared to the total activity of this radionuclide in the batch. Values above 95% are desirable, although not always achievable because a different label will yield a different biodistribution, resulting in a distortion of the image. The radiochemical purity can be determined by planar chromatography or electrophoresis combined with subsequent autoradiography. More popular are column chromatography methods:

  (a) Gel filtration of radiopharmaceuticals or size exclusion chromatography uses a three-dimensional mesh of porous beads in which smaller molecules can penetrate, whereas the large ones cannot. Hence, smaller molecules are retained longer, unlike for other forms of chromatography.

  (b) High-pressure liquid chromatography (HPLC) uses small absorbent particles of 5–10 μm as a stationary phase, tightly packed in thin columns. The large surface area for exchange and partitioning gives excellent resolution. HPLC is popular because it allows the separation of practically all impurities and degradation products and permits the development of stability-indicating assays. A reservoir of the mobile phase is pumped through a sample injector, and the chromatographic column passes in front of a γ-ray detector. Sample volumes of 10–50 μL are sufficient and help avoid nonlinear detector responses and excessive detector dead time effects [99].

  HPLC detection of radioactivity can be done simultaneously with ultraviolet (UV) absorption or refractive index (RI) measurements. This allows the simultaneous determination of the amount of isotopic carrier provided that a calibration of the UV absorption or the RI can initially be done with a sample of the carrier material. Thus, the specific activity can also be determined by such measurements [9].

  (c) Electrophoretic methods use the different migration velocities of charged solute species dissolved in an electrolyte under the influence of an electric field gradient in a rigid medium like paper or cellulose acetate.

- Traces of residual solvents must be quantified, which is frequently done by gas chromatography.
- Control of pH and osmolarity.
- Sterility can be ensured by conventional ultrafiltration techniques.
- Apyrogenicity can be confirmed by the *Limulus* amoebocyte lyase test. However, pyrogens are soluble and heat resistant and cannot be eliminated by filtration techniques or autoclave sterilisation. The only reliable technique is to ensure that radiopharmaceutical production starts and finishes pyrogen free [100] since pyrogens are products of microbial metabolism and degradation sterility is an important factor.

- Biodistribution tests are complex and time consuming but necessary in the development of new radiopharmaceuticals.

The exact methods to be applied may differ from one radiopharmaceutical to another, but the manufacturing and QC of radiopharmaceuticals has to follow the processes described in corresponding monographs of the pharmacopoeia published by the European Council [98] or of the national pharmacopoeias of the member states of the European Council. Only if none of these documents is available is the U.S. pharmacopoeia or a monograph of a third state acceptable. These documents provide full specifications for official preparations. Exemptions can be made for commercially available good manufacturing practice (GMP) labelling kits if they are applied according to the GMP-approved prescriptions of the supplier.

6.7 Quality Assurance and Good Manufacturing Practice

Radiopharmaceuticals, like any other pharmaceutical and medicinal products, have to be produced in compliance with GMP. The European Commission Directive 2003/94/EC [101] presented the principles and guidelines of GMP for medicinal products for human use. This directive defines GMP as "the part of the quality assurance system which ensures that products are consistently produced and controlled in accordance with the quality standards appropriate for their intended use." The manufacturer of medicinal products is obliged to ensure that manufacturing operations are carried out in accordance with GMP and with the specifications that are part of the authorizations for manufacturing and marketing. For this purpose, the manufacturer has to implement and run a pharmaceutical QA system that comprises all organised activities required to ensure that the product always fully meets the quality standards required for its intended use. The directive also calls for regular reviews of the manufacturing and QC procedures to incorporate if appropriate new technical and scientific developments to increase the quality and safety of the product. This means that GMP and QA are not static requirements but require continuous review and improvement. Compliance with GMP comprises the radiopharmaceutical production facility and its equipment, the personnel involved in the production, and all consumable and raw materials used in the production process.

The radiopharmaceutical production facility has to be properly designed, constructed, maintained, and monitored to suit the intended operations. Concerning the handling of radioactivity, the main aim is to prevent the escape of radioactivity. Hence, the radiopharmaceutical production facility has to be in depression with respect to the public environment. At the same time, the product has to be protected from the environment, which means that the production laboratory has to be kept in slight overpressure with respect to the area in which it is embedded. This necessitates keeping a complex cascade of pressure and depression in the facility. The air quality has to meet specific requirements in every compartment of the production facility. Operation procedures have to be defined concerning the use of protective clothing and the monitoring of radioactive and bacteriological contamination, and cleaning schedules have to be established.

The available personnel resources, like the number of staff and their educational background and professional experience, must ensure compliance with GMP. Personal responsibilities of staff and hierarchical relationships must be clearly documented. Training, especially on quality and safety-related issues, must be ongoing, well planned, and documented, including its verification and its effectiveness.

In practice, the whole production process, the QC steps, the maintenance and calibration of equipment, the training of personnel, as well as administrative procedures have to be fixed in standard operation procedures (SOPs). Clear records must be kept that allow verification that predefined specifications are met in every single step directly or indirectly relevant to the manufacture of the product in compliance with the specifications that are part of the manufacturing and marketing authorizations. All critical materials that enter the manufacturing process have to have quality records that allow complete retraceability from the final product down to all materials and substances used in its production. Moreover, the manufacturer is obliged to follow up the market, to establish complaint procedures, and to document non-conformity of the product or possible adverse effects in patients.

Guidelines for the implementation of GMP in the manufacture of radiopharmaceuticals can be found in *EudraLex* [102] (Vol. 4, Annex 3).

6.8 Regulatory Aspects

Radiopharmaceuticals have to be considered unsealed radioactive sources for use in the human body. Therefore, they are among the most regulated products because pharmaceutical legislation and legislation concerning the protection of patients, workers, and the public against hazards of ionising radiation overlap in this area.

On the European level, the legal framework is set by a series of directives, regulations, and decisions. Directives define a result that has to be achieved in each member state, but the member state may implement these directives by applying different combinations of primary legislation, statutory regulation, or administrative arrangements as long as the objectives can be fully met. Regulations are legally binding and directly applicable in all member states after they have been published in the *Official Journal of the European Community*. In many cases, guidelines define how the requirements of the legal framework can be met best. They are usually not legally binding, but they can be considered as representing a harmonised community position that, if followed, will facilitate obtaining authorisations for manufacturing or marketing of medicinal products.

Council Directive 97/43/Euratom [103] lays down basic measures for the radiation protection of persons undergoing medical examination or treatment, and Council Directive 96/29/Euratom [104] and amending directives set the basic safety standards for the health protection of the general public and workers against the dangers of ionising radiation. These directives require authorisations for production, purchase, handling, medical use, and administration of radioactive substances. Any exposure of a patient to ionising radiation must be duly justified, ensuring an acceptable risk-benefit balance. In addition, the health and safety of the personnel involved in the manufacturing, distribution, and transport as well as of the medical personnel handling the substance in a clinical setup have to be safeguarded. Moreover, the exposure of the public to chemical and radiation hazards must be kept within the legal limits, and all measures serving this purpose have to be reviewed from time to time concerning their appropriateness in the light of new technical and medical developments. These reviews have to include as well aspects of environmental protection and waste management. An environmental risk assessment (ERA) in accordance with Directive 2001/18/EC is mandatory [105]. Also, compliance with national and international regulations on the transport of dangerous materials is mandatory (cf. [106, 107]).

An overview of EU pharmaceutical legislation can be found on the home page of the European Commission, Directorate General Enterprise and Industry (see *EudraLex*) [102]. One of the key documents is Directive 2001/83/EC on the community code relating to medicinal products for human use [108]. This directive (and the earlier ones it replaced) obliges member states to operate a system of authorizations for the manufacture and marketing of medicinal products. According to Article 6 of Directive 2001/83/EC, a marketing authorization in accordance with regulations EC/726/2004 [109] and EC/1394/2007 [110] is obligatory for radionuclide generators, labelling kits, and radionuclide precursors as well as for radiopharmaceuticals. Regulation EEC 2309/93 [111] established a centralised EU-wide marketing authorization and the European Medicines Agency (EMEA), which releases centralised marketing authorizations. The assessment of the submissions to EMEA is undertaken by teams of experts drawn from national authorities coordinated by two members of the EU Committee for Human Medicinal Products (CHMP).

Directive 2001/83/EC specifies that a marketing authorisation is required prior to placing any medicinal product on the market and specifies content and structure of the dossier to be submitted when applying for marketing authorisation concerning the product characteristics, its manufacture, and medical indication. In this directive, procedures for obtaining a marketing authorisation and for the mutual recognition of national marketing authorisations in other member states are defined. Also, manufacture and processing of medicinal products are subject to prior authorisation specifying the medicinal products and pharmaceutical forms that are to be manufactured or imported and the place where they are to be manufactured or controlled. For obtaining a manufacturing authorisation for a specified medicinal product, the applicant has to prove that the facilities and technical equipment and the number and qualification of personnel are sufficient for the applicant to meet the required safety and quality standards. Article 48 obliges the holder of a manufacturing license to have at the holder's disposal, permanently and continuously, the service of at least one qualified person whose qualifications, responsibilities, and duties are specified explicitly. Article 47 obliges the manufacturer to implement and follow GMP. Moreover, labelling, packaging, classification, and distribution of medicinal products, advertising, and product information are regulated. An important point is the pharmacovigilance to ensure that competent authorities and users always have an updated overview of suspected adverse reactions of the product. Member states are obliged to supervise adherence to the directive by a system of repeated inspections.

Part III of the Annex of Directive 2001/83/EC deals with radiopharmaceuticals and explicitly requires that also the nuclear reactions involved in the radionuclide production shall be discussed when applying for market authorization. The application file shall include details on the identity of the radionuclide and its use, impurities, carriers, and data on specific activity. It is specified that the irradiated target material shall be considered as a starting material. The processing of the irradiated target material to extract and purify the desired radionuclide is considered as the first radiopharmaceutical production step. The radionuclide purity, the radiochemical purity, and the specific activity shall be described considering their impact on the biodistribution of the radiolabelled substance. Additional guidance was provided by the CHMP [112].

6.9 Radionuclide Availability and Prospectives

Medical imaging and therapy using radionuclides have become increasingly successful in the last 10–15 years. First, targeted radionuclide therapies obtained market approval and gave evidence of the enormous potential that radionuclide therapy holds in the treatment of cancer and other malignancies. However, nuclear medicine and radiopharmacy frequently are facing more and more problems with the global supply of radionuclides (e.g., [113]). Reactor-produced radionuclides were so far considered more readily available than cyclotron-produced ones. For this reason, 80% of the medical imaging procedures are still done with ^{99m}Tc-labelled compounds, and Zevalin and Bexxar are labelled with ^{90}Y and ^{131}I, respectively. Nevertheless, the lack of commercially available ^{90}Y and ^{177}Lu, approved for use in human medicinal products, is more and more perceived as an obstacle in the diffusion of new radionuclide therapies [27], whereas the lack of cyclotron-produced, mainly short-lived, radionuclides for therapy has been noted for several years (e.g., [60, 114]).

The worldwide supply with reactor-produced medical radionuclides relies on five research reactors (HFR in Petten, Netherlands; NRU in Chalk River, Canada; OSIRIS in Saclay, France; BR2 in Mol, Belgium; SAFARI-1 at Pelindaba, South Africa). These facilities range in age from 42 to 51 years. Their reliable operation is an increasing challenge, and many of them will be shut down within 5 to 6 years [115]. Only Australia's recently constructed research reactor OPAL at Lucas Heights is expected to start ^{99}Mo production soon, and the United States considers starting radionuclide production using the research reactor at the University of Missouri [11]. The project to construct two reactors dedicated to medical radionuclide production in Canada was cancelled in May 2008 [116].

As of October 2008, among the dozens of radiopharmaceuticals used in clinical practice only seven products were approved centrally within the European Union by a market authorisation issued by the EMEA [117]. The others have national market authorisations, some of which have been extended by mutual recognition agreements to other countries. Therefore, a clear assessment of the impact of the radionuclide shortage due to unexpected reactor shutdowns by national authorities and the European Commission is a complex exercise; hence, the conclusions drawn on a political level and on the level of the medical experts administering these radiopharmaceuticals are sometimes not in complete agreement, although converging [118].

In the short term, the shortage of ^{99}Mo supply for ^{99m}Tc generators is dealt with by prioritising more urgent clinical procedures on a hospital level or by using alternative radiopharmaceuticals. However, the application of a radiopharmaceutical for purposes not originally foreseen in the application for market authorisation requires a new and usually time-consuming application. Therefore, the European Union, member states, and EMEA on September 18, 2008, agreed on an ad hoc procedure allowing the national regulatory authorities to extend the licenses of marketing authorisation holders rapidly after having received further supportive data [119]. A further problem arises from the fact that the targets used in various reactors for the production of ^{99}Mo are not standardised and require different processing [71]. Since the target processing procedure and the location where processing takes place are part of the market authorisation, the processing of different targets irradiated in another reactor requires authorisation from regulatory authorities and an amendment to the Drug Master File. In addition, further transport authorisations are required for targets that are shipped from a different reactor to the processing site. Thus, the possibilities to react rapidly on the shortage are limited.

In view of the ageing reactors supplying the world with medical radionuclides, joint efforts are required to identify other research reactors with the technical and intellectual capacity to produce medical radionuclides [120, 121]. Also, possibilities to substitute reactor-produced radionuclides by accelerator-produced ones should be explored, especially if this could boost more efficient use of already installed accelerators. EMEA also calls for exploring alternatives to radiopharmaceuticals whenever new emerging technologies could allow this. But so far, unlike other imaging modalities, nuclear medicine procedures are uniquely capable of mapping physiological function and metabolic activity on a molecular level. Hence, the radionuclide supply problems may jeopardise the dynamics of the development in nuclear medicine but not the general trend.

Acknowledgements I thank F. Barberis Negra (GE Healthcare, Ispra) for helpful comments on the daily praxis of QC and GMP.

References

1. Nijsen, J.F.W., Krijger, G.C., van het Schip, A.D., The bright future of radionuclides for cancer therapy. Anti-Cancer Agents Med Chem. 7, 271–290 (2007)
2. Volkert, W.A., Hoffman, T., Therapeutical radiopharmaceuticals. Chem Rev. 99, 2269–2292 (1999)

3. Nuclear Technology Review 2007, IAEA/NTR/2007, Annex II -Radiopharmaceuticals: production and availability. IAEA, Vienna, Austria, pp. 60–71 (2007)
4. Ruhlmann, J., Oehr, P., Biersack, H.-J. (eds.), PET in Oncology, basics and clinical applications. Springer, Berlin, Heidelberg, New York etc. (1999)
5. IAEA-TECDOC-1340, Manual for reactor produced radioisotopes. International Atomic Energy Agency, Vienna, Austria (2003)
6. Kidd, L., Curies for patients. Nucl Eng Int 53, 26–32 (2008)
7. Technical Reports Series No. 465, Cyclotron produced radionuclides: principles and practice. International Atomic Energy Agency, Vienna, Austria (2008)
8. Stamm, H. and Gibson, P.N., Research with light ion cyclotrons in Europe, in: Proceedings of the 5th International Conference on Isotopes, 5ICI, Brussels, Belgium, April 25–29, 2005; Medimond S.r.l., Bologna, Italy, pp. 1–5 (2005)
9. Machulla, H.-J., Positron emitting radionuclides, in Sampson, C.B., ed., Textbook of radiopharmacy: theory and practice, 3rd ed. Gordon and Breach Science Publishers, Australia/Canada/China/France/Germany/India/Japan/Luxembourg/Malaysia/The Netherlands/Russia/Singapore/Switzerland, pp. 31–35 (1999)
10. Elliott, A.T., Radionuclide generators, in Sampson, C.B., ed., Textbook of radiopharmacy: theory and practice, 3rd ed. Gordon and Breach Science Publishers, Australia/Canada/China/France/Germany/India/Japan/Luxembourg/Malaysia/The Netherlands/Russia/Singapore/Switzerlandpp. 19–29 (1999)
11. Feder, T., US weighs entering radioisotope market. Phys Today. 61, 22–24 (2008)
12. Brans, B., Linden, O., Giammarile, F., Tennvall, J., Punt, C., Clinical applications of newer radionuclide therapies. Eur J Cancer. 42, 994–1003 (2006)
13. Stahl, A.R., Freudenberg, L., Bokisch, A., Jentzen, W., A novel view on dosimetry-related radionuclide therapy: presentation of a calculatory model and its implementation for radioiodine therapy of metastasized differentiated thyroid carcinoma. Eur J Nucl Med Mol Imaging. 36, 1147–1155 (2009)
14. Rubino, C., De Vathaire, F., Dottorini, M.E., Hall, P., Schvartz, C., Couette, J.E., Dondon, M.G., Abbas, M.T., Langlois, C., Schlumberger, M., Second primary malignancies in thyroid cancer patients. Br J Cancer. 89, 1638–1644 (2003)
15. Lam, M.G.E.H., de Klerk, J.M.H., van Rijk, P.P., Zonnenberg, B.A., Bone seeking radiopharmaceuticals for palliation of pain in cancer patients. Anti-Cancer Agents Med Chem. 7, 381–397 (2007)
16. Stigbrand, T., Carlsson, J. and Adams, G.P. (eds.), Targeted radionuclide tumor therapy – biological aspects. Springer Science+Business Media B.V., New York (2008)
17. Sampson, C.B., ed., Textbook of radiopharmacy: theory and practice, 3rd ed. Gordon and Breach Science Publishers, Australia/Canada/China/France/Germany/India/Japan/Luxembourg/Malaysia/The Netherlands/Russia/Singapore/Switzerland (1999)
18. Hoskin, P. (ed.), Radiotherapy in practice: radioisotope therapy, Oxford University Press, Oxford (2007)
19. O'Donoghue, J.A. and Wheldon, T.E., Targeted radiotherapy using Auger electron emitters. Phys Med Biol. 41, 1973–1992 (1996)
20. Wiseman, G.A., Witzig, T.E., Yttrium-90 (90Y) ibritumomab tiuxetan (Zevalin) induces longterm durable responses in patients with relapsed or refractory B-cell non-Hodgkin's lymphoma. Cancer Biother Radiopharm 20, 185–188 (2005)
21. Pohlman, B., Sweetenham, J., Macklis, R.M., Review of clinical radioimmunotherapy. Exp Rev Anticancer Ther 6, 445–461 (2006)
22. Davies, A.J., Radioimmunotherapy for B-cell lymphoma: Y90-ibritumomab tiuxetan and I131 tositumomab. Oncogene. 26, 3614–3628 (2007)
23. Morschhauser, F., Radford, J., Van Hoof, A., Vitolo, U., Soubeyran, P., Tilly, H. Huijgens, P.C., Kolstad, A., d'Amore, F., Diaz, M.G., Petrini, M., Sebban, C., Zinzani, P.L., van Oers, M.H.J., van Putten, W., Bischof-Delaloye, A., Rohatiner, A., Salles, G., Kuhlmann, J., Hagenbeek, A. Phase III trial of consolidation therapy with yttrium-90 -ibritumomab tiuxetan compared with no additional therapy after first remission in advanced follicular lymphoma. J Clin Oncol. 26, 5156–5164 (2008)

24. Reubi, J.C., Peptide receptors as molecular targets for cancer diagnosis and therapy. Endocr Rev. 24, 389–427 (2003)
25. Reubi, J.C., Schär, J.-C., Waser, B., Wenger, S., Heppeler, A., Schmitt, J.S., Mäcke, H. Affinity profiles for human somatostatine receptor subtypes SST1 – SST5 of somatostatine radiotracers selected for scintigraphic and radiotherapeutic use. Eur J Nucl Med. 27, 273–282 (2000)
26. Kwekkeboom, D.J., Müller-Brand, J., Paganelli, G., Anthony, L.B., Pauwels, S., Kvols, L.K., O'Dorisio, T.M., Valkema, R., Bodei, L., Chinol, M., Maecke, H., Krenning, E.P., Overview of results of peptide receptor radionuclide therapy with 3 radiolabelled somatostatin analogues. J Nucl Med. 46, 62S–66S (2005)
27. Prasad, V., Fetscher, S. and Baum, R.P., Changing role of somatostatin receptor targeted drugs in NET: Nuclear Medicine's view. J Pharm Pharm Sci. 10, 321s–337s (2007)
28. De Jong, M., Verwijnen, S.M., de Visser, M., Kwekkeboom, D.J., Valkema, R., Krenning, E.P. Peptides for radionuclide therapy, in Stigbrand, T., Carlsson, J., Adams, G.P., eds., Targeted radionuclide tumor therapy – biological aspects. Springer Science + Business Media, New York, pp. 117–144 (2008)
29. Becherer, A., Szabó, M., Karanikas, G., Wunderbaldinger, P., Angelberger, P., Raderer, M., Kurtaran, A., Dudczak, R., Kletter, K., Imaging of advanced neuroendocrine tumours with 18F-FDOPA PET. J Nucl Med. 45, 1161–1167 (2004)
30. Prasad, V., Ambrosini, V., Alavi, A., Fanti, S., Baum, R.P., PET/CT in neuroendocrine tumours: evaluation of receptor status and metabolism. PET Clin. 2, 351–375 (2007)
31. Antunes, P., Ginj, M., Zhang, H., Wasser, B., Baum, R.P., Reubi, J.C., Maecke, H., Are radiogallium-labelled DOTA-conjugated somatostatin analogues superior to those labelled with other radiometals? Eur J Nucl Med Mol Imaging. 34, 982–993 (2007)
32. De Jong, M., Kwekkeboom, D., Valkema, R., Krenning, E.P., Tumour therapy with radiolabelled peptides: current status and future directions. Dig Liver Dis. 36, S48–S54 (2004)
33. Stigbrand, T., Eriksson, D., Riklund, K., Johansson, L. Targeting tumours with radiolabelled antibodies, in Stigbrand, T., Carlsson, J., Adams, G.P., eds., Targeted radionuclide tumor therapy – biological aspects. Springer Science + Business Media, New York, pp. 59–76 (2008)
34. Goldenberg, D.M., Sharkey, R.M., Paganelli, G., Barbet, J., Chatal, J.F., Antibody pretargeting advances cancer radioimmunodetection and radioimmunotherapy. J Clin Oncol. 24, 823–834 (2006)
35. Boerman, O.C., Van Schaijk, F.G., Oyen, W.J.G., Corstens, F.H.M., Pretargeted radioimmunotherapy of cancer: progress step by step. J Nucl Med. 44, 400–411 (2003)
36. Cremonesi, M., Ferrari, M., Chinol, M., Bartolomei, M., Stabin, M.G., Sacco, E., Fiorenza, M., Tosi, G., Paganelli, G., Dosimetry in radionuclide therapies with 90Y-conjugates: the IEO experience. Quart J Nucl Med. 44, 325–332 (2000)
37. Stoldt, H.S., Aftab, F., Chinol, M., Paganelli, G., Luca, F., Testori, A., Geraghty, J.G., Pretargeting strategies for radio-immunoguided tumour localization and therapy. Eur J Cancer. Part A, 186–192 (1997)
38. Faraji, A.H. and Wipf, P., Nanoparticles in cellular drug delivery. Bioorg Med Chem. 17, 2950–2962 (2009)
39. Rahman, W.N., Bishara N., Ackerly, T., He, C.F., Jackson P., Wong, C., Davidson, R., Geso, M., Gold nanoparticles: clinical nanomedicine, radiation oncology – enhancement of radiation effects by gold nanoparticles for superficial radiation therapy. Nanomed Nanotechnol Biol Med. 5, 136–142 (2009)
40. Hamoudeh, M., Kamleh, M.A., Diab, R., Fessi, H., Radionuclides delivery systems for nuclear imaging and radiotherapy of cancer. Adv Drug Deliv Rev. 60, 1329–1346 (2008)
41. Bouchat, V., Nuttens, V.E., Lucas, S., Michiels, C., Masereel, B., Féron, O., Gallez, B., Borght, T.V., Radioimmunotherapy with radioactive nanoparticles: First results of dosimetry for vascularized and necrosed solid tumours. Med Phys. 34, 4504–4513 (2007)
42. Torchilin, V.P., Targeted pharmaceutical nanocarriers for cancer therapy and imaging. AAPS J. 9, E128–E147 (2007)

43. DeNardo, S.J., DeNardo, G.L., Miers, L.A., Natarajan, A., Foreman, A.R., Gruettner, C., Adamson, G.N., Ivkov, R., Development of tumor targeting bioprobes (111In-chimeric L6 monoclonal antibody nanoparticles) for alternating magnetic field cancer therapy. Clin Cancer Res. 11, 7087s–7092s (2005)
44. Hayashi, K., Moriya, M., Sakamoto, W., Yogo, T. Chemoselective synthesis of folic acid-functionalized magnetite nanoparticles via click chemistry for magnetic hyperthermia. Chem Mater. 21, 1318–1325 (2009)
45. Tolmachev, V. Choice of radionuclides and radiolabelling techniques, in Stigbrand, T., Carlsson, J., Adams, G.P., eds., Targeted radionuclide tumor therapy – biological aspects. Springer Science + Business Media, New York, pp. 145–174 (2008)
46. Carlsson, J., Forssell Aronsson, E., Hietala, S.-O., Stigbrand, T., Tennvall, J., Tumour therapy with radionuclides: assessment of progress and problems. Radiother Oncol. 66, 107–117 (2003)
47. Bethge, K., Kraft, G., Kreisler, P., Walter, G., Medical Applications of Nuclear Physics, Springer-Verlag, Berlin, Heidelberg, New York, Hong Kong, London, Milan, Paris, Tokyo (2004)
48. Laforest, R. and Liu, X., Image quality with non-standard nuclides in PET. Quart J Nucl Med. 52, 151–158 (2008)
49. National Nuclear Data Center, Brookhaven National Laboratory; Sonzogni, A. (database manager and Web programming), http://www.nndc.bnl.gov/nudat2
50. Erdi, Y.E., The use of PET for radiotherapy. Curr Med Imaging Rev. 3, 3–16 (2007)
51. Modak, A., Breath tests with ^{13}C substrates. J Breath Res. 3, p.1 (2009)
52. Goddard, A.F. and Logan, R.P.H., Review article: urea breath tests for detecting Helicobacter pylori. Aliment Pharmacol Ther. 11, 641–649 (1997)
53. Uusijärvi, H., Bernhardt, P., Ericsson, T., Forssel-Aronsson, E., Dosimetric characterization of radionuclides for systemic tumour therapy: influence of particle range, photon emission, and subcellular distribution. Med Phys. 33, 3260–3269 (2006)
54. Wheldon, T.E. and O'Donoghue, The radiobiology of targeted radiotherapy. Int J Radiat Biol. 58, 1–21 (1990)
55. O'Donoghue, J.A., Bardiès, M., Wheldon, T.E., Relationships between tumor size and curability for uniformly targeted therapy with beta-emitting radionuclides. J Nucl Med. 36, 1902–1909 (1995)
56. Sgouros, G. High-LET-Emitting radionuclides for cancer therapy, in Stigbrand, T., Carlsson, J., Adams, G.P., eds., Targeted radionuclide tumor therapy – biological aspects. Springer Science + Business Media, New York, pp. 175–180 (2008)
57. Lundqvist, H., Stenerlöw, B., Gedda, L., The Auger effect in molecular therapy, in Stigbrand, T., Carlsson, J., Adams, G.P., eds., Targeted radionuclide tumor therapy – biological aspects. Springer Science + Business Media, New York, pp. 195–214 (2008)
58. Lechner, A., Blaickner, M., Gianolini, S., Poljanc, K., Aiginger, H., Georg, D. Targeted radionuclide therapy: theoretical study of the relationship between tumour control probability and tumour radius for a 32P/33P radionuclide cocktail. Phys Med Biol. 53, 1961–1974 (2008)
59. Pratt, B., Evans, S. Physics principles in the clinical use of radioisotopes, in Hoskin, P., ed., Radiotherapy in practice: radioisotope therapy, Oxford University Press, Oxford, pp. 1–8 (2007)
60. Weinreich, R., Molecular radiotherapy with ^{211}At, in Amaldi, U., Larsson, B., Lemoigne, Y., eds., Advances in hadrontherapy, Elsevier Science, Amsterdam, pp. 359–382 (1997)
61. Dearling, J.L.J., Pedley, R.B. Antinbody directed radioisotope therapy, in Hoskin, P., ed., Radiotherapy in practice: radioisotope therapy; Oxford University Press, Oxford pp. 9–45 (2007)
62. Howell, R.W., Goddu, S.M., Rao, D.V., Proliferation and the advantage of longer-lived radionuclides in radioimmunotherapy. Med Phys. 25, 37–42 (1998)
63. Flynn, A.A., Green, A.J., Pedley, R.B., et al. A model-based approach for the optimization of radioimmunotherapy through antibody design and radionuclide selection. Cancer. 94, 1249–1257 (2002)

64. Couturier, O., Supiot, S., Degraef-Mougin, M., Faivre-Chauvet, A., Carlier, T., Chatal, J.F., Davodeau, F., Cherel, M. Cancer radioimmunotherapy with alpha-emitting nuclides. Eur J Nucl Med Mol Imaging. 32, 601–614 (2005)
65. Morgenstern, A., Apostolidis, C., Bruchertseifer, F., Capote, R., Gouder, T., Simonelli, F., Sin, M., Abbas, K. Cross-sections of the reaction 232Th(p,3n)230 Pa for production of 230U for targeted alpha therapy. Appl Radiat Isot. 66, 1275–1280 (2008)
66. Boll, R.A., Malkemus, D., Mirzadeh, S., Production of actinium-225 for alpha particle mediated radioimmunotherapy. Appl Radiat Isot. 62, 667–679 (2005)
67. Henriksen, G., Hoff, P., Larsen, R.H., Evaluation of potential chelating agents for radium. Appl Radiat Isot. 56, 667–671 (2002)
68. de Goeij, J.J.M., Bonardi, M.L., How to define the concepts specific activity, radioactive concentration, carrier, carrier-free and no-carrier-added? J Radioanal Nucl Chem. 263, 13–18 (2005)
69. Bonardi, M.L., de Goeij, J.J.M., How do we ascertain specific activities in no-carrier-added radionuclide preparations? J Radioanal Nucl Chem. 263, 87–92 (2005)
70. Mani, R.S. Reactor production of radionuclides for generators. Radiochim Acta. 41, 103–110 (1987)
71. Saey, P.R.J. The influence of radiopharmaceutical isotope production on the global radioxenon background. J Environ Radioact. 100, 396–406 (2009)
72. Zhernosekov, K.P., Filosofov, D.V., Baum, R.P., Aschoff, P., Bihl, H., Razbash, A.A., Jahn, M., Jennewein, M., Rösch, F., Processing of generator produced 68 Ga for medical application. J Nucl Med. 48, 1741–1748 (2007)
73. Pillai, M.R.A., Chakraborty, S., Das, T., Venkatesh, M., Ramamoorthy, N. Production logistics of ^{177}Lu for radionuclide therapy. Appl Radiat Isot. 59, 109–118 (2003)
74. Lebedev, N.A., Novgorodov, A.F., Misiak, R., Borckmann, J., Rösch, F., Radiochemical separation of no-carrier-added ^{177}Lu as produced via the [176]Yb(n, γ)[177]Lu process. Appl Radiat Isot. 53, 421–425 (2000)
75. Canella, L., Bonardi, M.L., Groppi, F., Persico, E., Zona, C., Menapace, E., Alfassi, Z.B., Chinol, M., Papi, S., Tosi, G., Accurate determination of the half-life and radionuclidic purity of reactor produced [177g]Lu ([177m]Lu) for metabolic radiotherapy. J Radioanalyt Nucl Chem. 276, 813–818 (2008)
76. Palmer, M., Basic mechanisms of radiolabelling, in Sampson, C.B., ed., Textbook of radiopharmacy: theory and practice, 3rd ed. Gordon and Breach Science Publishers, Australia/Canada/China/France/Germany/India/Japan/Luxembourg/Malaysia/The Netherlands/Russia/Singapore/Switzerland, pp. 57–62 (1999)
77. Lindegren, S., Bäck, T., Jensen, H.J. Dry-destillation of astatine from irradiated bismuth targets: a time-saving procedure with high recovery yields. Appl Radiat Isot. 55, 157–160 (2001)
78. Groppi, F., Bonardi, M.L., Birattari, C., Menapace, E., Abbas, K., Holzwarth, U., Alfarano, A., Morzenti, S., Zona, C., Alfassi, Z.B., Optimization study of α-cyclotron production of At-211/Po211g for high-LET metabolic radiotherapy purposes. Appl Radiat Isot. 63, 621–631 (2005)
79. Holzwarth, U., Schaaff, P., Abbas, K., Schaub, W., Hansen-Ilzhöfer, Maier, K. Production of miniaturized [72]Se/[72]As positron generators for applications in materials science, in Proceedings 17th International Conference on Cyclotrons and Their Applications (CYCLOTRONS 2004), RIKEN, KEK, 18–22 October 2004, Tokyo, Japan (2004)
80. Skarnemark, G., Solvent extraction and ion exchange in radiochemistry, in Vértes, A., Nagy, S., Klencsár, Z., eds., Handbook of nuclear chemistry, Vol. 5, Chapter 7, Kluwer Academic, Dordrecht, Boston, London (2004)
81. Zona, C., Bonardi, M.L., Groppi, F., Morzenti, S., Canella, L., Persico, E., Menapace, E., Alfassi, Z.B., Abbas, K., Holzwarth, U., Gibson, N., Wet-chemistry method for the separation of no-carrier added [211]At/[211g]Po from [209]Bi target irradiated by alpha-beam in cyclotron. J Radioanal Nucl Chem. 276, 819–824 (2008)
82. Abbas, K., Kozempel, J., Bonardi, M., Groppi, F., Alfarano, A., Holzwarth, U., Simonelli, F., Hofman, H., Horstmann, W., Menapace, E., Lešetický, L., Gibson, N., Cyclotron production of ^{64}Cu by deuteron irradiation of [64]Zn. Appl Radiat Isot. 64, 1001–1005 (2006)

83. Kozempel, J., Abbas, K., Simonelli, F., Zampese, M., Holzwarth, U., Gibson, N., and Lešetický, L., A novel method for n.c.a. ^{64}Cu production by the ^{64}Zn(d, 2p)^{64}Cu reaction and dual ion-exchange column chromatography. Radiochim Acta. 95, 75–80 (2007)
84. Hess, E., Takács, S., Scholten, B., Tárkányi, F., Coenen, H.H., Qaim, S.M., Excitation function of the ^{18}O(p, n)^{18}F nuclear reaction from threshold up to 30 MeV. Radiochim Acta 89, 357–362 (2001)
85. Lucignani, G., PET Imaging with hypoxia tracers: a must in radiation therapy. Eur J Nucl Med Mol Imaging. 35, 838–842 (2008)
86. Stigbrand, T., Eriksson, D., Riklund, K., Johansson, L., Therapeutically used targeted antigens in radioimmunotherapy, in Stigbrand, T., Carlsson, J., Adams, G.P. eds., Targeted radionuclide tumor therapy – biological aspects. Springer Science + Business Media, New York, pp. 13–23 (2008)
87. Bryan, R.A., Jiang, Z., Huang, X., Morgenstern, A., Bruchertseifer, F., Sellers, R., Casadevall, A., Dadachova, E., Radioimmunotherapy is effective against high-inoculum cryptococcus neoformans infection in mice and does not select for radiation-resistant cryptococcal cells. Antimicrob Agents Chemother. 53, 1679–1682 (2009)
88. Dadachova, E., Patel, M.C., Toussi, S., Apostolidis, C., Morgenstern, A., Brechbiel, M.W., Gorny, M.K., Zolla-pazner, S., Casadevall, A., Goldstein, H., Targeted killing of virally infected cells by radiolabelled antibodies to viral proteins. PloS Med. 3, 2094–2103 (2006)
89. Goetz, C., Riva, P., Poepperl, G., Gildehaus, F.J., Hischa, A., Tatsch, K., Reulen, H-J., Locoregional radioimmunotherapy in selected patients with malignant glioma: experiences, side effects and survival times. J Neuro-Oncol. 62, 321–328 (2003)
90. Wilbur, D.S., Radiohalogenation of proteins: an overview of radionuclides, labeling methods, and reagents for conjugate binding. Bioconjug Chem. 3, 433–470 (1992)
91. Adam, W.J., Wilbur, D.S., Radiohalogens for imaging and therapy. Chem Soc Rev. 34, 153–163 (2005)
92. Bakker, W.H., Krenning, E.P, Reubi, J.C., Breeman, W.A., Setyono-Han, B., de Jong, M., Kooij, P.P., Bruns, C., van Hagen, P.M., Marbach, P., In vivo application of [^{111}In-DTPA-D-Phe^1]octreotide for detection of somatostatin receptor-positive tumours in rats. Life Sci. 49, 1593–1601 (1991)
93. Nowotnik, D.P., Verbruggen, A.M., Practical and physicochemical aspects of the preparation of ^{99m}Tc-labelled radiopharmaceuticals, in Sampson, C.B., ed., Textbook of radiopharmacy: theory and practice, 3rd ed. Gordon and Breach Science Publishers, Australia/Canada/China/France/Germany/India/Japan/Luxembourg/Malaysia/The Netherlands/Russia/Singapore/Switzerland, pp. 37–55 (1999)
94. DeNardo, G.L., DeNardo, S.J., Wessels, B.W., Kukis, D.L., Miyao, N., Yuan, A., ^{131}I-Lym-1 in mice implanted with human Burkitts's lymphoma (Raij) tumors: loss of tumour specificity due to radiolysis. Cancer Biother Radiopharm. 15, 547–560 (2000)
95. Wahl, R.L., Wissing, J., del Rosario, R., Zasadny, K.R., Inhibition of autoradiolysis of radiolabelled monoclonal antibodies by cryopreservation. J Nucl Med. 31, 84–89 (1990)
96. Chakrabarti, M.C., Le, N., Paik, C.H., De Graff, W.G., Carrasquillo, J.A., Prevention of radiolysis of monoclonal antibody during labeling. J Nucl Med. 37, 1384–1388 (1996)
97. Liu, S., Ellars, C.E., Edwards, D.S., Ascorbic acid: useful as a buffer agent and radiolytic stabilizer for metalloradiopharmaceuticals. Bioconjug Chem. 14, 1052–1056 (2003)
98. European Pharmacopoeia, ed. by Council of Europe, European Directorate for the Quality of Medicines (EDQM), Strasburg, France, 6th Edition 6.6 (2009)
99. Theobald, A.E., Quality control of radiopharmaceuticals, in Sampson, C.B., ed., Textbook of radiopharmacy: theory and practice, 3rd ed. Gordon and Breach Science Publishers, Australia/Canada/China/France/Germany/India/Japan/Luxembourg/Malaysia/The Netherlands/Russia/Singapore/Switzerland, pp. 145–186 (1999)
100. Keeling, D.H., Adverse reactions and untoward events associated with the use of radiopharmapharmaceuticals, in Sampson, C.B., ed., Textbook of radiopharmacy: theory and practice, 3rd ed. Gordon and Breach Science Publishers, Australia/Canada/China/France/Germany/India/Japan/Luxembourg/Malaysia/The Netherlands/Russia/Singapore/Switzerland, pp. 431–445 (1999)

101. Commission Directive 2003/94/EC of 8 October 2003 laying down the principles and guidelines of good manufacturing practice in respect of medicinal products for human use and investigational medicinal products for human use. Official J. L 262, 14/10/2003, pp. 22–26
102. EudraLex, The rules governing medicinal products in the European Union, Volume 4, EU guidelines to good manufacturing practice, medicinal products for human and veterinarian use; European Commission, Enterprise and Industry Directorate General (2008), http://ec.europa.eu/enterprise/pharmaceuticals/eudralex/eudralexen.htm
103. Council Directive 97/43/Euratom on health protection of individuals against the dangers of ionizing radiation in relation to medical exposure. Official J. L 180, 09/07/1997, pp. 22–27
104. Council Directive 96/29/Euratom laying down basic safety standards for the protection of the health of workers and the general public against the dangers arising from ionizing radiation. Official J. L 159, 29/06/1996, pp. 1–114
105. Directive 2001/18/EC of the European Parliament and of the council of 12 March 2001 on the deliberate release into the environment of genetically modified microorganisms and repealing Council Directive 90/220/EEC. Official J. L 106, 17/4/2001, pp. 1–39
106. Training Course Series No. 1, G.E., Safe transport of radioactive material, 3rd ed., International Atomic Energy Agency, Vienna, Austria (2002)
107. IAEA Safety Standards Series No. TS-G-1.5, Compliance assurance for the safe transport of radioactive material – safety guide, International Atomic Energy Agency, Vienna, Austria (2009)
108. Directive 2001/83/EC of the European Parliament and the council of 6 November 2001 on the community code relating to medicinal products for human use. Official J. L 311, 28/11/2001, pp. 67–128 (consolidated version of 30/12/2008 and amended directives and regulations)
109. Regulation (EC) No 726/2004 of the European Parliament and the Council of 31 March 2004 laying down Community procedures for the authorization and supervision of medicinal products for human and veterinary use and establishing a European Medicines Agency. Official J. L, 30/4/2004, pp. 1–33 (consolidated version 20/4/2009)
110. Regulation (EC) No/1394/2007 of the European Parliament and of the council of 13 November 2007 on advanced therapy medicinal products and amending Directive 2001/83/EC and Regulation EC/726/2004, Official J. L 324, 10/12/2007, pp. 121–137
111. Council Regulation (EEC) No 2309/93 of 22 July 1993 laying down Community procedures for the authorization and supervision of medicinal products for human and veterinary use and establishing a European Agency for the Evaluation of Medicinal Products. Official J. L 214, 24/8/1993, pp. 1–21
112. Guideline of radiopharmaceuticals, Committee for Human Medicinal Products (CHMP), European Medicines Agency (2007), EMEA/CHMP/QWP/306970/2007
113. Perkins, A., Hilson, A., Hall, J., Global shortage of medical isotopes threatens nuclear medicine services. Brit Med J. 357, a1577 (2008)
114. Tolmachev, V., Carlsson, J., Lundqvist, H., A limiting factor for the progress of radionuclide-based cancer diagnostics and therapy. Acta Oncol. 43, 264–275 (2004)
115. Nuclear Technology Review 2004, Annex I: research reactors, pp. 28–37. International Atomic Energy Agency, Vienna, Austria (2004)
116. International Atomic Energy Agency, Addressing the global shortage of beneficial radiation sources, Staff Report, 4 November 2008, http://www.iaea.org/NewsCenter/News/2008/resreactors.html
117. European Medicines Agency, Report to the European Commission on the supply shortage of radiopharmaceuticals (status as of 24 October 2008); London, 5 March 2009, EMEA/51183/2009
118. European Association of Nuclear Medicine; Press release 02/12/2008: Long term measures needed
119. European Medicines Agency, Public statement on the current shortage of radiopharmaceuticals in the European Union, London, 01/10/2008, EMEA/501698/2008
120. Preliminary draft report of the SNM Isotope Availability Task Group, Society of Nuclear Medicine, February 2009
121. Gould, P., Medical radioisotope shortage reaches crisis level. Nature. 460, 312–313 (2009)

Chapter 7
Research and Development of New Radiopharmaceuticals

Gerhard Holl

7.1 Introduction

The goal of research and development (R&D) for pharmaceutical companies is to find new products, to demonstrate that these are efficacious and safe, and to get them approved by the health authorities, thus making them available for broad use by the medical community. This comprises a long chain of activities, including identification of medical needs, extensive research work, preclinical preparation for administration of the new drug to humans, and a series of clinical studies. Although this chapter focuses on R&D of radiopharmaceuticals (RPs), general aspects of R&D of (nonradioactive) pharmaceutical products are also presented. Intravenous injection is always considered as the route of administration of RPs.

There are major differences between R&D of diagnostic RPs and of therapeutic RPs. This applies particularly to the investigation of the effects of ionizing radiation (which can be measured in individual patients after administration of therapeutic agents but can only be estimated as risks in the case of diagnostic RPs) and the strategies of elaborating doses. In this chapter, all statements refer to diagnostic RPs unless otherwise noted.

The focus is on molecular imaging (MI). In this field, mainly such tracers are considered which reversibly bind to defined molecular structures. Special focus is on positron emission tomography (PET). The assessment of radiation risk is a crucial step in R&D of RPs. It requires significant contribution by experts in medical physics, and discussion is presented here in greater detail.

The general aspects in sections 7.4 and 7.5 are illustrated by examples that mainly refer to one clinical field: the diagnosis of dementia. If detailed data are given for a special tracer, they refer to BAY94-9172, which is currently in clinical development at Bayer Schering Pharma. This tracer binds reversibly to deposits of amyloid β, a protein that is found in the brains of patients with Alzheimer's disease (AD).

G. Holl
Bayer Schering Pharma, Clinical Development Molecular Imaging, Berlin, Germany
e-mail: Gerhard.holl@bayerhealthcare.com

M.C. Cantone and C. Hoeschen (eds.), *Radiation Physics for Nuclear Medicine*,
DOI 10.1007/978-3-642-11327-7_7, © Springer-Verlag Berlin Heidelberg 2011

7.2 Medical Needs

As a prerequisite for success, a new product must fulfill a medical need; that is, it must perform in a relevant medical field better than currently possible. In the field of diagnostic RPs, the desired improvement over existing techniques and products usually is better diagnostic performance. Improvement in safety and tolerability of the chemical entity of the tracer plays a minor role since most current RPs do not exhibit safety issues. However, reduction of exposure of patients to ionizing radiation can be considered an additional goal.

The following examples of medical needs refer to improved diagnostic performance in three disease groups that affect large numbers of patients: cancer, cardiovascular diseases, and dementia.

- *Cancer imaging*: 2-Deoxy-2-[^{18}F]fluoro-D-glucose ([^{18}F]-FDG) is the most frequently used PET tracer worldwide and shows excellent sensitivity in many cancer types. However, [^{18}F]-FDG cannot reliably distinguish between cancerous lesions and sites of inflammation. Erroneous classification of inflammation as cancer metastasis and vice versa can lead to wrong therapeutic decisions. New tracers need to have the same excellent sensitivity as [^{18}F]-FDG in a broad range of different cancer types but should be more specific for cancer.
- *Cardiovascular imaging*: Atherosclerosis is the leading cause of morbidity and mortality in industrialized countries. Although there are many diagnostic tools, diagnosis remains suboptimal. Radiological and nuclear medicine imaging techniques help to detect narrowing of blood vessels and the reduction of blood flow in the heart regions supplied by the narrowed arteries. These techniques are not able to identify inflamed and vulnerable atherosclerotic plaques that may be present in vessel segments that are not significantly stenosed. However, such non-stenosing plaques are responsible for many acute coronary events, including cardiac infarctions. There is the chance that MI of vulnerable plaques will better predict cardiac infarction than the currently available tools, and that infarction can be better prevented by early initiation of adequate treatment. Several molecular markers of vulnerable plaques have been identified (e.g., markers of cell death in the inflamed lesions) [1].
- *Diagnosis of dementia*: Dementia is the fourth leading cause of death in the Western world. The most common cause of dementia is AD. Definitive diagnosis of dementia relies on neuropathological examination of the brain. Examination of brain samples, however, requires the performance of a brain biopsy, which is a highly invasive procedure. For that reason, neuropathological examination is usually only performed after the patient has died.
  The diagnosis of AD in living patients requires time-consuming clinical assessment by dementia specialists. Many cases, however, remain unclear. With magnetic resonance imaging (MRI) and PET imaging with [^{18}F]-FDG, brain areas with major structural alterations respectively with reduced glucose metabolism can be recognized. Both imaging methods, however, have difficulties in distinguishing the slight alterations seen in patients at early disease stages from nonpathological age-related brain alterations.

Improvement of diagnostic accuracy and facilitation in making the diagnosis thus is a medical need in the field of dementia. MI with tracers that are specific for the neuropathological hallmarks of dementias are expected to provide information that is close to what otherwise can be obtained only by neuropathology.

7.3   Overview of R&D Process

A schematic overview of the steps of R&D with RPs is given in Fig. 7.1.

By research, a series of compounds is made available that potentially can be used as tracers for imaging. The most promising molecule (lead compound) is selected. Before a first administration of the new tracer in humans, two major tasks need to be accomplished:

1. *Chemistry, manufacturing, and control (CMC) work.* Different chemical substances, including the radiolabeled tracer molecule, are combined to obtain the "formulation" of the final product for injection. The formulation of RPs usually contains solvents (such as water and ethanol), buffers for pH adjustment, and other constituents. The CMC unit establishes manufacturing of the formulation and of all constituents according to good manufacturing practice (GMP). The stability of the product is investigated, and the maximum time period between manufacturing and injection is specified.

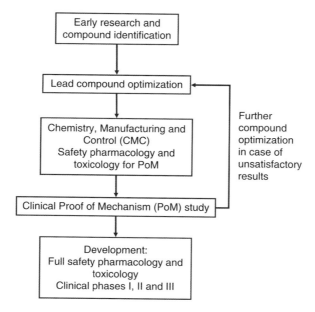

Fig. 7.1   Overview of research and development of radiopharmaceuticals

2. *Safety pharmacology and toxicology* testing of the product by performing in vitro tests and animal studies. Here, R&D of RPs can benefit from special regulations, which were recently issued by the U.S. Food and Drug Administration (FDA) [2] and the European Medicines Agency (EMEA) [3]. These regulations recommend reduction of safety pharmacology and toxicology programs provided that a number of "microdosing" conditions are fulfilled: In studies with a single administration of the tracer, the maximum dose to be applied to humans must be $100 \mu g$ or less and must be less than 1/100th of the dose calculated to yield a pharmacological effect (based on in vitro and animal data). Regarding the clinical study, the number of study participants must be low and the study needs to be exploratory.

The microdosing regulations can be applied for many RPs since tracer mass doses are often in the low-microgram range, and no effects are observed in preclinical studies with the required doses. The demand for small size of the clinical study can also be met because nuclear medicine imaging can provide conclusive information on tracer biodistribution by investigating a small number of healthy volunteers or patients.

The clinical *proof of mechanism (PoM) study* is the first study in which the new tracer is applied to humans. In many cases, the study population will consist of a small number of patients with a certain disease and a small number of healthy volunteers as controls. The main objective of a PoM study is to evaluate in humans whether the tracer shows uptake due to specific binding to the molecular target. Imaging therefore is focused on the body regions and anatomical structures where, on the one hand, the molecular target is expressed with high density and where, on the other hand, differences in the density of the target are expected between healthy and diseased persons. In addition to the principle proof of in vivo binding in humans, first information is obtained whether the differences found between the two groups of study participants might be large enough to allow discrimination between subjects who have the disease and those who do not. As in all clinical studies, safety and tolerability of the new product are continuously monitored. In addition to imaging of the target regions, whole-body imaging is performed at several time points to obtain first data on the human biodistribution for an estimation of radiation dosimetry data. If the new tracer does not fulfill expectations, further compound optimization will be performed.

Clinical development: If results of the PoM study are promising, the clinical development is started. Microdosing regulations do not apply for clinical development. Therefore, the full set of safety pharmacology and toxicology investigations as required for other pharmaceutical products needs to be performed.

Phase I studies are mostly performed in healthy volunteers. The objective is the assessment of pharmacokinetics (PK), including metabolism of the tracer molecule, and of safety and tolerability of the formulation. In the case of RPs, one further important objective is to obtain data on radiation dosimetry in a larger group of subjects.

Phase II studies always include patients in defined disease states in whom abnormal imaging results are expected. Subjects in whom normal imaging results are expected may also be included, for example, healthy volunteers or patients with diseases that do not alter the molecular target. The objectives are as follows:

- Identification of the optimum activity dose. For reasons of radiation protection, the administered activity for diagnostic purposes should be as low as reasonably achievable (ALARA principle). On the other hand, the administered activity must be sufficient to provide image quality that is good enough for making the diagnosis. Within certain limits, the applied activity can be reduced without loss of image quality by using a longer image acquisition time. Too lengthy acquisition periods, however, are not well tolerated by patients and may lead to movement artifacts, which in turn make correct interpretation of images impossible. Thus, reasonable compromises need to be made in selecting the optimal activity.
- Identification of optimum imaging windows (times recommended for image acquisition). For this purpose, the scan protocols include several acquisition periods, often ranging from a starting point soon after tracer injection up to two or three half-lives of the radioactive decay. After evaluation of the phase II images, a more simple imaging protocol is recommended for the subsequent phase III studies.
- Elaboration of rules how to make the diagnosis by evaluating the image data.
- Generation of hypotheses on the diagnostic performance of the tracer.

Phase III studies are the final clinical development step. They are confirmatory studies. This means that a certain diagnostic performance, predefined on the basis of the phase II results (usually sensitivity and specificity in the detection of disease), must be reached.

There usually only one standard activity dose is applied, and the imaging protocol is kept as simple as possible, desirably consisting of not more than one acquisition period. For image data evaluation, the rules as elaborated in the phase II studies are applied. The procedures applied in phase III studies will be recommended later for the use of the marketed product.

One goal of the phase III studies is to demonstrate that the use of the new tracer is not limited to a few highly specialized institutions. Phase III studies therefore are conducted in many different nuclear medicine departments, in different countries and regions, and including institutions with scanners from different equipment manufacturers.

7.4 Special Aspects of Research

7.4.1 Targets for Molecular Imaging

The principle of targeting for MI is shown in Fig. 7.2. Knowledge of the deviations from normal physiology or biochemistry in the tissues affected by the disease is a prerequisite.

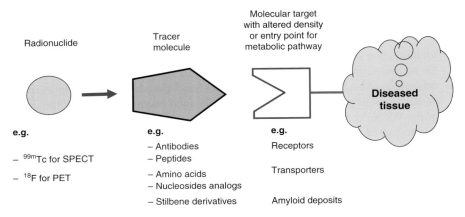

Fig. 7.2 Principle of targeting in molecular imaging

One imaging principle depends on the altered density of certain receptors or other molecular structures to which radiolabeled tracers can bind. The tracer will show a pathologically high accumulation if the density of the molecular structure is higher in the diseased than in the healthy tissue (e.g., binding to a tumor marker). If the density of the target structure is reduced as a consequence of the disease, the tracer may show pathologically reduced uptake (e.g., reduced uptake of radiolabeled raclopride in the striatum in some degenerative brain diseases).

A second approach is imaging of the metabolism. Here, the radiolabeled tracer is a substrate for a special metabolic pathway. In oncology, cancer can be visualized by pathologically high accumulation of radiolabeled amino acids (indicating elevated protein synthesis), of radiolabeled nucleosides (indicating elevated synthesis of DNA), or of $[^{18}F]$-FDG (indicating elevated glucose consumption). Also in metabolic imaging, some diseases can be recognized by pathologically reduced accumulation of the radiolabeled tracer (e.g., reduced uptake of 6-$[^{18}F]$fluoro-L-dopa in the striatum of patients with Parkinson's disease).

The last example in Fig. 7.2 refers to imaging in dementia. The molecular target chosen is amyloid β, a marker of AD. Amyloid β is not a receptor with a physiological function. However, its presence is closely associated with AD, and it has a defined chemical structure that makes it suitable for molecular targeting. The tracer molecules chosen in Fig. 7.2 are radiolabeled stilbene derivatives that bind to amyloid β.

The selection of molecular targets as the basis of an R&D program for MI is performed by a cooperation between scientists who have knowledge about the biology and biochemistry of healthy and diseased human tissues and clinicians who are aware of the medical needs. In the following, some aspects of target selection for imaging are discussed. MI of AD is given as a special example.

The chosen molecular target (or metabolic process) should

1. allow the diagnosis to be made at an early stage of the disease, at a time when diagnosis by other means is difficult or even impossible. In the case of neurodegenerative diseases, this aspect is of special relevance when assuming that neuroprotective drugs will be available in the future. Then, reliable diagnosis at an early disease stage is absolutely needed to enable start of neuroprotective treatment before significant numbers of nerve cells have died.
2. the promise high sensitivity in disease detection. The first prerequisite for this is presence of a molecular alteration in a high percentage of patients. In the ideal case, it should be present in 100% of the patients with the respective disease. As a second prerequisite, the target structure must allow high uptake of the radiotracer. One condition for this is that the target structure appears with a high density in the tissue and is easily accessible for the tracer.
3. Promise high specificity in disease detection.

In the case of AD, many abnormalities on the cellular and molecular levels are known, including several nonspecific processes, such as inflammation and cell death. Amyloid β plaques are present with high density at the beginning of the disease and are – according to current knowledge of the disease process – developing years before the first clinical symptoms of dementia become obvious. Therefore, amyloid β plaques are considered suitable targets for sensitive detection of AD at early disease stages.

Regarding specificity, amyloid β imaging can distinguish AD from types of dementia that are not associated with amyloid β (e.g., frontotemporal dementia [FTD]) (see Fig. 7.4). However, deposits of amyloid β can also be present in a fraction of patients with two less-common types of dementia: dementia with Lewy bodies (DLB) and vascular dementia (VaD). As a consequence, discrimination between AD and these two types of dementia can possibly not be made reliably. Thus, if the main role is seen in the diagnosis of diseases, in the considered case the discrimination of AD versus other types of dementia, the specificity of amyloid β imaging may not be perfect. This situation is not uncommon in MI since the alterations detected on the molecular level are not always matched 1:1 to a specific disease entity.

However, beyond the capability to confirm or exclude specific diseases (e.g., to exclude AD in case of a normal scan), an important question is whether a proof of presence or absence of the molecular alteration and thus of a defined pathology as such is of clinical value. In the case of dementia, therapeutic strategies against amyloid β deposits are under clinical investigation by several pharmaceutical companies. If one of these strategies will be successful, amyloid β imaging should be an important tool in identifying the patients who can benefit from the new therapy. The detection or exclusion of a defined pathology, in this case the distinction between patients with and without brain amyloid β load, by means of MI then might be of higher importance than the classification of a patient's condition according to the current disease definitions. This point of view is also relevant in selecting targets for MI.

7.4.2 Research for New MI Tracers

The first condition in selecting tracer molecules is binding of the tracer molecules to the molecular target with high affinity and selectivity. However, several other pharmaceutical conditions must be met to obtain the final tracer molecule, such as having the lipophilicity (octanol–water partition coefficient) in a defined range, the possibility of radiolabeling the compound without changing its structure too much (otherwise, binding properties may be changed), maintaining adequate chemical and metabolic stability, and the possibility of manufacturing the compound at a reasonable cost.

One effective approach is to start with a molecule that is known as a binder (e.g., a dye that is used by pathologists to stain amyloid plaques) and to modify its structure to achieve the pharmaceutical properties mentioned. In this process, the chemists use known structure–activity relationships to improve certain features of the compound.

A second approach is to screen available molecules for their capability to bind to the chosen molecular target. This can be done efficiently by high-throughput screening (HTS) of large libraries of candidate compounds: The purified target molecules or cells exhibiting the target molecules are filled into the wells of a microtiter plate. Interaction of one of the screened compounds with the target is detected by competitive binding assays (e.g., displacement of a labeled reference compound from the target).

Other techniques can generate completely new molecular structures that bind to the chosen targets. Methods that were successfully applied in the field of RPs are the phage display technique for antibody-based tracers and the SELEX (systematic evolution of ligands by exponential enrichment) method for tracers consisting of aptamers (short strands of nucleic acids).

7.4.3 Preclinical Characterization of New Tracers

7.4.3.1 Binding Studies

The affinity, the strength of binding between the tracer molecule and the target, is one of the key parameters of a new tracer. Affinity is best characterized by the equilibrium dissociation constant K_d. This constant is equal, in a binding experiment, to the concentration of the tracer molecule at which half of the receptors are occupied by the tracer molecule under equilibrium conditions. The smaller the K_d is, the stronger the binding will be. K_d values of radiotracers should be in the low nanomole-per-liter range.

The selectivity of the tracer molecule is investigated by studying its binding to a panel of other human receptors and transporters. Significant binding to structures other than the targeted one may indicate specificity issues in imaging.

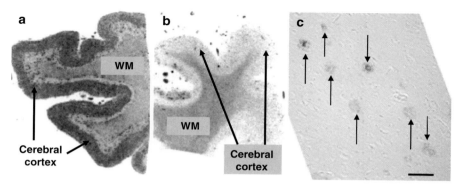

Fig. 7.3 Autoradiography of brain tissue sections from a patient with Alzheimer's disease (AD) who died and from a person who died without signs of dementia. (**a**) Postmortem brain tissue section from a patient with AD. Autoradiography after incubation with the amyloid β-targeting tracer BAY94-9172 shows strong signals of the radiotracer in the outer rim zone of the brain section (cerebral cortex) due to specific binding. Less-intense signal is seen in the white matter (WM) due to nonspecific binding. (**b**) Brain tissue section from a person who died without signs of dementia. Autoradiography again shows some signal due to unspecific binding in the WM but no relevant signal in the cerebral cortex. (**c**) Enlarged view of an area of the cerebral cortex of the patient with AD whose autoradiography is shown in (**a**). Immunohistochemistry with the antiamyloid β monoclonal antibody 11F6 (Bayer Schering Pharma) shows presence of amyloid β plaques (*arrows*). Horizontal bar: 50 μm. (Bayer Schering Pharma, TRG Diagnostic Imaging, A. Thiele and S. Krause)

Binding of a tracer molecule to the target in human tissue samples can be analyzed by autoradiography. The brain tissue sections shown in Fig. 7.3 were from a patient who died with the diagnosis of AD and from a person who had no signs of dementia at the time of death. The sections were incubated with the ^{18}F-labeled amyloid β targeting tracer BAY94-9172. After a washing step, binding of the tracer to the target is visualized using a phosphor imaging plate for detection of the autoradiographic signal. The main difference between the two images is the strong signal in the cerebral cortex of the patient with AD and the absence of a relevant signal in the cortex of the person who was not suffering from dementia. In both images, minor binding is seen in the white matter (WM), a structure that is not primarily involved in the deposition of amyloid β in AD. This signal is considered to be due to nonspecific binding to molecular structures unknown until now. Figure 7.3c shows the application of a further method used in research, immunohistochemistry: The same sections of the cerebral cortex as used for the autoradiography were stained with an antibody that binds selectively to amyloid β. Presence of amyloid β deposits was demonstrated only in the cortex of the patient with AD and not in the cortex of the nondemented person. This finding in immunohistochemistry supports the conclusion that it is the binding of the tracer to amyloid β deposits that makes the difference between the two autoradiography studies shown in Fig. 7.3a and 7.3b.

The two methods were also applied to obtain information on the selectivity of the tracer BAY94-9172 that can be expected in diagnosing dementias associated with amyloid β, such as AD versus other types of dementia. Figure 7.4a shows the

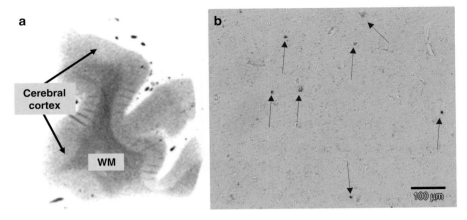

Fig. 7.4 Autoradiography of a brain tissue section from a patient who died with frontotemporal dementia (FTD). (**a**): Postmortem brain tissue section from a patient with FTD. Autoradiography after incubation with the amyloid β-targeting tracer BAY94-9172 shows the same pattern as seen in the sections from a person who died without signs of dementia (Fig. 7.3b). (**b**) Enlarged view of an area of the cerebral cortex of the FTD patient. Immunohistochemistry with the monoclonal anti-hyperphosphorylated tau antibody AT8 shows presence of depositions of hyperphosphorylated tau (*arrows*). Horizontal bar: 100 μm. (Bayer Schering Pharma, TRG Diagnostic Imaging, S. Krause and A. Thiele)

autoradiography of a brain tissue section from a patient who died with the diagnosis of FTD, a disease generally not associated with amyloid β deposits. No specific binding of the tracer to the cerebral cortex could be detected. Immunohistochemistry of the same brain section using an antibody specific for hyperphosphorylated tau revealed the presence of this pathological marker of FTD (Fig. 7.4b). These findings confirmed that the tracer does not bind to hyperphosphorylated tau and is therefore useful for making a differential diagnosis between AD and FTD.

7.4.3.2 Animal Studies for Radiation Dosimetry

In the development of a new RP, the first estimate of the expected radiation dose to humans is obtained from animal studies, usually performed in mice. These data are required prior to a first study in humans. For principles and methods of internal radionuclide dosimetry, see the discussion in the assessment of radiation risk of section 7.5.

The first step is the quantitative determination of activity in a large set of organs at several time points after intravenous injection. Imaging is not considered accurate enough for this purpose in small animals. Therefore, animals are sacrificed at pre-defined time points, and tissues are removed for measurement of activity in a well counter. The activity excreted in urine and feces is also quantitatively measured.

For extrapolation of animal data to humans, the data measured in the various source organs usually are scaled prior to calculation of the cumulated activity $\tilde{A}$ in humans. The choice of the most adequate scaling method is a matter of discussion. One method is organ and body mass scaling according to Kirschner et al. [4]. If the percentage of injected activity in an organ of the animal is known, the percentage injected activity in the human organ is obtained by multiplication with the factor

$$F_a = (\text{body weight/organ weight})_{\text{animal}} \cdot (\text{organ weight/body weight})_{\text{human}} . \quad (7.1)$$

A second method is scaling for biological equivalent times according to Boxenbaum [5]. If a time constant is found for the description of the time activity curve (TAC) of an animal organ, the appropriate time constant for humans is obtained by multiplication with the factor

$$F_b = (\text{body weight}_{\text{human}} / \text{body weight}_{\text{animal}})^{0.25} . \quad (7.2)$$

For the ^{18}F-labeled tracer BAY94-9172, the results of different approaches (no scaling of the mouse data, scaling only for body mass, scaling only for biological time, or scaling for both) were retrospectively compared to the dosimetry data measured in humans. The greatest level of agreement was obtained with the combination of scaling for body mass and biological time [6].

In the dosimetry of diagnostic RPs, the sole purpose of measuring the activity distribution in animals is to obtain data for estimates of the radiation doses to humans, whereas the radiation doses to animal organs are not of interest. In the case of therapeutic RPs, the situation is different. Determination of radiation doses to healthy organs of the animal and to pathological structures, such as xenograft tumors (see section "Imaging Xenograft Models"), and evaluation of radiation-induced organ damage and tumor shrinking provide valuable information about the capability of a product to treat cancer. The determination of radiation doses to animal organs can be difficult since S-values (e.g., for the mouse) are published for a few radionuclides and for some selected organs only [7].

7.4.3.3 Imaging Studies in Animals

Determination of tracer distribution in animals by means of imaging has advantages over the method described in Sect. 7.4.3.2 on animal studies for radiation dosimetry, which requires sacrificing animals at many postinjection time points: With imaging, postinjection time activity curves of various tissues can be determined in the same animal. This reduces the effects of interanimal variability and reduces the number of animals required for a given study. Furthermore, repeated studies in the same animal are possible. For many purposes, the information obtained by imaging is sufficient.

Prior to the development of small-animal imaging systems, tracer distribution could be observed only in larger animals by using clinical scanners. A widespread application of PET with larger animals was brain imaging in monkeys, including modelling of tracer kinetics.

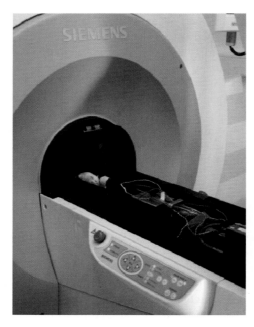

Fig. 7.5 State-of-the-art preclinical imaging equipment: positron emission tomographic/computed tomographic (PET/CT) system for animal studies. (Bayer Schering Pharma, TRG Diagnostic Imaging)

Over the past decade, dedicated small-animal systems for single-photon emission computed tomography (SPECT) [8] and PET [9] were developed, some of them combined with computed tomography (CT) systems. PET systems can achieve a spatial resolution in the range of 1–2 mm. Small-animal SPECT systems use pinhole collimation and can reach even better resolution. A modern PET/CT system for animal studies is shown in Fig. 7.5.

Imaging Genetically Engineered Animal Models

An advantage of small animals (mice and rats) over larger ones is the lower cost for keeping of the animals. On the other hand, the mouse is the animal of most interest because of the techniques available for genetic manipulations in this species. It is possible to inactivate ("knock out") genes or to insert new ones to obtain animals that mimic human diseases. By these means, certain tissues of the animals may exhibit molecular structures that otherwise can only be observed in patients suffering from that disease. Thus, a new tracer targeting these structures can be investigated by imaging such animal models. For AD, several models have been created that exhibit brain pathology similar to that observed in patients [10].

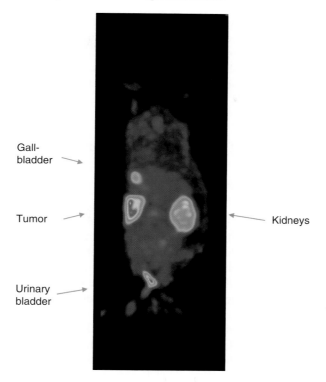

Gall-
bladder

Tumor

Kidneys

Urinary
bladder

Fig. 7.6 Preclinical imaging studies. Visualization of tracer distribution in a tumor-bearing mouse model by single-photon emission computed tomography (SPECT). The tracer under investigation is targeting a marker of neoangiogenesis. Strong accumulation of radioactivity is seen in the tumor. Apart from the tumor, the excretory pathways can be recognized (in the chosen projection, the two kidneys of the animal are superimposed). (Bayer Schering Pharma, TRG Diagnostic Imaging, D. Berndorff)

Imaging Xenograft Models

Xenografts are grafts in which the donor and the recipient are of different species. For investigation of tracers for cancer imaging, cells from a human tumor cell line can be transplanted (e.g., under the skin of a mouse or into a mouse organ). The animals are immunocompromised and therefore do not reject the human cells. After a couple of weeks, the tumor develops to a diameter that is large enough for imaging.

The properties of human xenograft tumors in animals are considered to be similar to those of the same tumor type in patients. As an example, the degree of neoangiogenesis (the formation of new blood vessels) should be similar. In the SPECT scan shown in Fig. 7.6, a strong uptake by a mouse tumor xenograft is seen when administering a tracer that is targeting a molecular marker of neoangiogenesis. This finding in the xenograft model suggests that the tracer might also show high uptake when imaging the same tumor type in humans.

Furthermore, Fig. 7.6 shows the hepatic excretion (via liver and gallbladder) and the renal excretion (via kidneys and urinary bladder) of activity. The low activity in all body regions except for those with excretory organs suggests that imaging in humans may also show low background activity. Thus, important additional information about the tracer is obtained by this imaging study performed at an early research stage.

7.5   Special Aspects of Clinical Development

7.5.1   Good Clinical Practice

Good clinical practice (GCP) is an international ethical and scientific quality standard for designing, conducting, recording and reporting studies that involve the participation of humans. The basis of GCP is the ethical principles developed by the World Medical Association (WMA) http://www.wma.net/en/30publications/10policies/b3/17c.pdf. This declaration has undergone several revisions, the most recent in October 2008. Adherence to GCP is an absolute obligation in any kind of clinical studies.

The International Conference on Harmonization (ICH), as an international body, defines unified standards of GCP for the European Union, Japan, the United States and other countries; these standards are in the form of guidelines, which governments can transpose into national regulations [11]. For pharmaceutical companies, the ICH guidelines facilitate global development because the requirements for proving safety and efficacy of new products are harmonized in most of the major markets.

There are two main aspects of GCP: The first is protection of rights of trial participants and their safety and well-being. The second aspect is the assurance that clinical studies are planned in a scientifically sound way and that the results are credible and accurate. Quality assurance therefore is one cornerstone of GCP. Performance of clinical studies in accordance with GCP requires in particular the following documents, activities, and measures, which all need to be defined by standard operating procedures (SOPs):

- *Investigator's brochure*: The brochure contains all relevant data about the new tracer and the formulation as known at the time of the clinical study.
- *Clinical study protocol*: The protocol contains all relevant information about a particular study.
- *Case report form*: This form is a paper sheet or an electronic system for documentation of study data for individual study participants as collected by the investigator.
- *Patient information form and informed consent form*: Patients and healthy volunteers are informed about the purpose of the study and about benefit and risks. In particular, they are informed that their participation is absolutely voluntary, and that they can withdraw their consent at any time without giving reasons. Consent needs to be given in written form.

Further requirements are insurance coverage for the study participants; approval of the clinical studies by ethical committees and health authorities (after review of the documents mentioned); anonymization of all data given to the sponsor (including removal of patients' and volunteers' names from all image material); archiving of all study data; and the nomination of persons responsible for quality assurance.

Information for Study Participants About Radiation Risks

In studies with RPs, the study participants must be informed about the risk due to the ionizing radiation applied. This risk can be expressed best by the effective dose (ED; see chapters on assessment of radiation risk). For laypersons, however, this information is not sufficient. Two types of additional information can be provided: The first is the comparison of the ED caused by the experimental tracer to the EDs caused by frequently used diagnostic procedures with which the patients and volunteers are familiar: nuclear medicine procedures such as mycardial perfusion or [18F]-FDG scans and X-ray examinations, including CT studies.

The second way is to convert the ED caused by the experimental tracer to the equivalent time in months or years to obtain the same ED from the natural background radiation [12]. For the latter, average values are taken for the country where the study participant lives. In Germany, for example, the average annual radiation exposure due to natural radiation is 2.1 mSv [13]. Assuming that the ED of the experimental RP is 4.0 mSv, the study participant is informed that the administration of the RP will cause a radiation dose that is comparable to the radiation dose he or she receives due to the natural radiation in a period of $4.0/2.1 = 1.9$ years.

7.5.2   Safety

In the application of RPs, two different types of potential risks to the patients need to be considered: the risks due to the chemical substances injected and the risks due to the application of ionizing radiation.

7.5.2.1   Compound-Related Safety

The chemical substances injected are the constituents of the formulation, that is, the radiolabeled tracer molecule and the additives. Additives are selected from a list of substances that are approved for use in medications (e.g., by the European Pharmacopoeia). The safety profiles of these substances are well known from their use in other drugs. Substances causing significant adverse reactions can be excluded. If special features of the formulation suggest that there might be minor effects (e.g., some pain at the injection site due to low pH), the occurrence of these symptoms will be specially monitored in the clinical studies.

Regarding the tracer molecules, their mass doses are low, often in the microgram range. This, together with the fact that in the majority of cases the patient receives only one single application, is the explanation why diagnostic RPs are a safe class of products. The low rate of adverse reactions caused by marketed RPs was documented by several large surveys (e.g., by the European Association of Nuclear Medicine [14]).

However, when investigating new molecules, tracer-related changes in safety parameters and side effects cannot be excluded a priori, even with very low mass doses. Examples have been reported in the literature. An article by Chianelli et al. [15], for example, described laboratory parameter changes caused by radiolabeled interleukin 2 with tracer masses as low as 40 μg. Another report attributed cases of serious adverse events (AEs), including two cases of death [16], to an antibody product that was applied with tracer mass doses in the range of 75–125 μg. To be able to detect potential risks early, safety monitoring in clinical studies with RPs is not much different from that in studies with nonradioactive drugs.

Safety monitoring is most intensive during the first clinical studies with a new tracer. If the earlier studies confirmed good safety of the tracer, the duration of the observation period (the time interval during which the study participant's condition is surveyed) can be decreased in the later part of the development; as well, the number of assessments points and of safety variables can be decreased.

In the individual study participant, the measurements for evaluation of safety start with the assessment of the status at baseline (i.e., prior to the administration of the new tracer). Then, the same assessments are made at a number of defined time points after injection up to the end of the observation period. These measurements include

- *Physical examination*: This follows a standardized schedule (e.g., general appearance, heart, lungs, etc.).
- *Vital signs*: These include measurement of blood pressure and pulse rate. In early studies, body temperature, respiration rate, and oxygen saturation are also measured.
- *Clinical laboratory*: Measurement is made of parameters of chemistry (e.g., electrolytes, liver enzymes), of hematology (counts of blood cells and platelets), and of blood coagulation. Furthermore, the urine is analyzed.
- *Electrocardiograms (ECG)*: These are recorded and assessed for rhythm abnormalities and quantitative parameters.
- *Adverse events (AEs)*: An AE can be any unfavorable and unintended sign or symptom or disease that appears after administration of the tracer. Any AEs have to be documented irrespective of whether they are considered to be related to the application of the study tracer. AEs are to be described according to a number of categories, such as the intensity, the assumed causal relationship to the tracer molecule, and others.

Studies with antibodies of mouse origin: Administration of tracers consisting of murine monoclonal antibodies may trigger the patient's immune response to form human antimouse antibodies (HAMAs). This is a natural phenomenon and does not cause issues at the time of the first contact with murine proteins. In further administrations of murine material to a patient with HAMA, however, the interaction between the patient's HAMA and the injected tracer can lead to side effects such as fever, chills, and rashes. One special task in the clinical testing of such tracers is to assess the formation of HAMAs. For this purpose, HAMA levels in the patients' blood are determined after 2–4 weeks and again after several months. Furthermore, the safety of repeated administration is investigated.

7.5.2.2   Assessment of Radiation Risk

The effects of irradiation can be classified as either deterministic or probabilistic (stochastic).

Deterministic effects occur after larger doses of radiation to tissues. They are characterized by a threshold dose below which these effects will not be observed. Deterministic effects increase in severity as the radiation dose is increased. Examples of symptoms due to deterministic effects are the depression of cell division in the bone marrow or diarrhea due to irradiation of the gastrointestinal tract.

In clinical development of therapeutic RPs, the occurrence of deterministic effects needs to be closely surveyed. With many therapeutic RPs, the effects caused in one organ are "dose limiting"; that is, the administered activity cannot be increased further because serious damage of this organ would occur with higher activities.

With diagnostic RPs, the radiation doses are too low to reach the thresholds for deterministic effects in any organ. Therefore, a survey for occurrence of deterministic effects is not part of the safety monitoring in clinical studies with diagnostic RPs.

Stochastic effects refer to the occurrence of malignant diseases and of heritable effects. Such effects are all-or-none responses for which the probability of occurrence, but not the severity, is regarded as a function of the radiation dose. The assessment of the risk of stochastic effects is based on the linear nonthreshold (LNT) model. This model assumes that, in the low-dose range, radiation doses greater than zero will increase the risk of cancer or heritable diseases in a simple proportional manner. This means that stochastic effects need to be considered in any administration of ionizing radiation, irrespective of the magnitude of the radiation doses. Therefore, radiation dosimetry is a cornerstone of the development of any new RPs.

Principles of Internal Radionuclide Radiation Dosimetry

After injection of the RP into the vein, the activity distribution is not uniform but shows regions with elevated activity concentrations. Usual sources are organs (e.g., thyroid or liver) or the contents of organs (e.g., of bowel segments). The activity concentrations in the regions change over time due to biological mechanisms and because of the radioactive decay of the radionuclide.

It is the task of internal radionuclide radiation dosimetry to determine the radiation energy absorbed by the various human tissues on the basis of this time-dependent distribution pattern and to estimate the radiation risk. The methodology now widely used was developed by the Medical Internal Radiation Dose (MIRD) committee [17] of the U.S. Society of Nuclear Medicine and is generally referred to as the MIRD formalism.

Quantities used

The *cumulated activity* $\tilde{A}$ is the time integral of activity in a given region and has the dimension of activity × time (Bq × s). $\tilde{A}$ is equal to the sum of all nuclear transitions in the region during a given time interval.

The absorbed dose D is defined as the energy that is absorbed from ionizing radiation per unit mass of tissue and has the unit Joule per kilogram, which is equal to gray (Gy). In larger regions (organs), the mean absorbed dose $\bar{D}$ is of interest.

For *calculation of absorbed doses from cumulated activities*, the MIRD scheme defines source regions and target regions. In source regions, the course of activity over time is known, and the cumulated activity $\tilde{A}$ thus can be determined. The target region is the region for which the mean absorbed dose $\bar{D}$ is to be determined. The radiation absorbed dose to a given target region r_k (e.g., the liver) arises from the cumulated activity $\tilde{A}_k$ in this specified region r_k. However, additional radiation absorbed doses to the liver are caused by the cumulated activities $\tilde{A}_h$ in other regions r_h of the body due to the penetrating radiation that is emitted from the activity in the other regions and that is absorbed by the target region r_k.

The mean absorbed dose $\bar{D}_k$ in a given target region r_k and the cumulated activity $\tilde{A}_h$ in one source region r_h (including the cumulated activity $\tilde{A}_k$ in the target region r_k) are linked by the S value $S(r_k \leftarrow r_h)$:

$$\bar{D}_k = S\,(r_k \leftarrow r_h) \times \tilde{A}_h. \tag{7.3}$$

If one can estimate the cumulated activity $\tilde{A}_h$ for all relevant source organs r_h, the mean absorbed dose $\bar{D}_k$ for one defined target organ r_k can be estimated as

$$\bar{D}_k = \sum_h \left[S\,(r_k \leftarrow r_h) \times \tilde{A}_h \right]. \tag{7.4}$$

The S values depend on the source region and the target region considered, on the radionuclide, and on the model used (e.g., adult male or female or children at different ages). The S values for relevant organs, large numbers of radionuclides, and several models were developed and published by the International Commission on Radiological Protection (ICRP) and the MIRD committee and are integrated in computerized versions of the MIRD scheme.

The equivalent dose H is derived from the absorbed dose by multiplication with the dimensionless radiation weighting factor w_R. Since w_R is equal to 1 for gamma rays, X-rays, electrons, and positrons, the numerical value of H is identical to the D value for most RPs. The unit of H is also joule per kilogram but is called sievert (Sv).

The dimensionless tissue-weighting factor w_T describes the tissue-specific relative radiosensitivity for expressing cancer or genetic defects. The effective dose ED is obtained by multiplying the equivalent dose H_T of a tissue with the appropriate tissue-weighting factor w_T and summing the products for all specified tissues and organs of the body

$$ED = \sum_T [w_T \times H_T].\tag{7.5}$$

The *ED* represents a single-value estimate of the overall stochastic risk (i.e., the risk of cancer and genetic defects) of an irradiation applied to a human subject. In diagnostic RPs, *ED* is considered the most relevant quantity for describing such risks. The unit of *ED* is also sievert.

Weighting factors for calculation of the *ED* were published in ICRP Publication 60 [18]. Revised weighting factors are found in ICRP Publication 103 [19].

To get input data for the MIRD calculations, sources of activity within the body must be identified and quantified at various time points. Furthermore, the excretion of activity from the body via the main pathways should be investigated by collection of urine and feces. All measurements must provide quantitative results in terms of absolute activity (becquerel [Bq]) or fraction of administered activity. An overview of techniques was provided in MIRD Pamphlet 16 [20].

Imaging Techniques

The distribution of the activity in the body is determined by sequential whole-body imaging. With single-photon emitting radionuclides, the classical method for quantification is conjugate view counting, usually made by simultaneous anterior and posterior planar imaging with a double-head camera and calculation of the geometric mean of the counts. Various techniques were introduced to correct for attenuation, scatter, and background activity. Since planar imaging shows some limitations (e.g., in separating activity from overlapping organs), quantification by SPECT imaging was proposed [21].

With positron-emitting radionuclides, activity distributions in the body are acquired by whole-body PET. Quantitative PET imaging is generally thought to be more accurate than quantification by planar imaging and SPECT.

Whole-Body Imaging Protocols

During time periods with rapid changes in activities, acquisitions should be closely spaced. Usually, fast changes appear early after injection (peak uptakes and early washout). Measurement of activities at later time points should be performed until three times the effective half-life (a parameter describing disappearance of activity from the body due to the combined effects of radioactive decay and excretion).

For radiation dosimetry of the ^{18}F-labeled tracer BAY94-9172, nine whole-body acquisitions were performed, the first starting immediately after tracer injection, the last starting 6 h after injection, with the entire set of acquisitions thus covering more than three times the physical half-life of ^{18}F. For the first seven acquisitions with starting times up to about 2 h, the volunteers had to lay still on the scanner

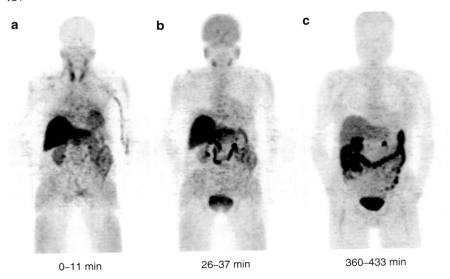

Fig. 7.7 Sequential whole-body imaging for radiation dosimetry. Whole-body images made with the ^{18}F-labeled tracer BAY94-9172 in a healthy volunteer (maximum intensity projections). The acquisition periods are indicated under the images. Images were selected from a series of nine acquisitions with starting times ranging from immediately after injection (**a**) to 6 h after injection (**c**). (Images courtesy of University Hospital Leipzig, Department of Nuclear Medicine. Director and chairman: Prof. Dr. Osama Sabri, MD, PhD)

table. They could then leave the scanner room and came back for the eighth acquisition starting 4 h after injection and again for the ninth acquisition starting 6 h after injection [22]. Figure 7.7 displays three selected whole-body images with starting times of zero, 26 min and 6 h after injection. Immediate uptake is seen in the liver (Fig. 7.7a), then the activity is eliminated via the gallbladder into the intestines (Fig. 7.7b and c), indicating the hepatobiliary pathway as one route of elimination from the body. In addition, part of the activity is eliminated via the kidneys and the urinary bladder. Figure 7.8 shows, for three selected organs of the hepatobiliary pathway (liver, gallbladder, upper large intestine) the observed activity as percentage of the activity administered at all nine time points. Corresponding to the chosen whole-body acquisition protocol, the early period after tracer injection is documented with higher time resolution (seven data points within 2 h) than the later parts (data points at 4 h and 6 h).

Evaluation of Whole-Body Image Data and Dose Calculations

By analyzing the whole-body images made at various time points, regions with significant accumulation of activity are identified. They are related by clinical investigators to source organs.

With planar conjugate view imaging, the procedures described in MIRD Pamphlet 16 [20] can be applied to quantify the activity in source organs.

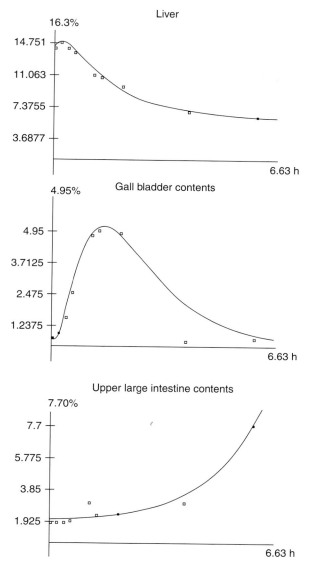

Fig. 7.8 Time activity curves (TACs) for determination of the cumulated activity $\tilde{A}$: TACs for the liver, the gallbladder contents, and the contents of the upper large intestine after administration of the ^{18}F-labeled tracer BAY 94-9172 in a healthy volunteer. Data points were obtained from a series of nine whole-body acquisitions with starting times between zero and 6 h after injection. Activities are expressed as percentage of the activity injected and are corrected for radioactive decay. Curves were exponentially fitted using the EXM module of the software OLINDA/EXM. The TACs reflect the hepatobiliary pathway of elimination of activity. They are the basis for calculation of the cumulated activities in the respective organs. (Images courtesy of University Hospital Leipzig, Department of Nuclear Medicine. Director and chairman: Prof. Dr. Osama Sabri, MD, PhD)

In the case of PET imaging, volumes of interest (VOIs) can be generated that cover each source organ. Summing up the activity in the VOI yields the activity in that organ at the time point considered for the individual subject studied. As a methodological variant, average standardized uptake values (SUVs; see Sect. 7.5.4.4) can be measured for the source organ at each time point. By use of the known mass of an organ in a "reference man" [23], the mean SUV can be converted to percentage of injected activity at each time point. This procedure provides an estimate of the activity present in the organ of a reference man using the tracer under investigation at that time point [24].

The result of this quantification is a time series of activity values for each source organ, usually expressed as percentage of A_0. The next step is the calculation of the cumulated activity $\tilde{A}$ (the time integral of activity) per organ. Several methods can be used to perform the integration. The software OLINDA/EXM [25] offers the option to enter the observed activities per source organ and time point and fits them to one or several exponential terms (Fig. 7.8). The software then integrates the fitted TACs and uses the integrals for dose calculations according to the MIRD scheme.

The most relevant results obtained from application of the MIRD formalism are the radiation doses to all target organs that are available in the chosen phantom and the ED. These data are essential information to investigators and are part of any prescribing information once the product is approved.

7.5.3 Pharmacokinetics

Pharmacokinetics (PK) studies are performed to determine the fate of substances that are administered to living organisms. PK data must be generated for each product that is intended for use in humans. This also includes RPs, despite their often extremely low mass doses.

The different areas of PK analysis are described by absorption, distribution, metabolism, and excretion, commonly referred to as ADME scheme. Absorption is the process by which the compound enters the body. Apart from a few exceptions with other routes of application (oral administration, inhalation), RPs are injected into the vein, so that absorption does not need a specific investigation. Once in the bloodstream, distribution takes place to various body regions, and metabolic reactions act on the compound, thereby transforming the parent compound into metabolites. Excretion is the elimination of the compound from the body.

In the case of diagnostic RPs, the activity in various organs can be determined with good accuracy by the methods of imaging and activity quantification outlined in the previous sections. When the fixation of the radionuclide to the tracer molecule is highly stable and the effects of the metabolism on the tracer molecule are negligible, the measured activity can be related quantitatively to the parent compound and can serve for a precise determination of the distribution and excretion of the tracer molecule.

With many tracers, however, significant metabolism takes places. Then, only a fraction of the measured activity is fixed to the parent molecule, whereas other fractions are fixed to metabolites or may represent the unbound radionuclide (e.g., fluoride or pertechnetate in the case of ^{18}F- or ^{99m}Tc-labeled tracers, respectively). Since these fractions cannot be distinguished from each other by imaging techniques, the imaging data cannot be used to describe the PK of the tracer molecule. Thus, as this is the case with nonradioactive drugs, quantification of the tracer parent molecule and its metabolites is possible in biological samples only (i.e., blood and excreta), for which adequate bioanalytical methods can be applied.

PK studies are part of the phase I program. They require collection of multiple blood samples and of excreta over an extended period of time. The basic PK parameters for RPs are the same as for nonradioactive drugs, such as the maximum concentration of the tracer molecule in plasma, the area under the plasma concentration time curve, and others [26].

7.5.4 Diagnostic Efficacy

7.5.4.1 Mathematical Modeling of PET Tracer Kinetics in the Brain

For modeling the kinetics of a PET tracer, studies are performed in the early part of clinical development in a small number of patients and healthy persons. The goal of modeling is to analyze the relationship between the data that can be measured by scanning and the physiological or biochemical parameters that determine the binding or the metabolism of the tracer. The most frequently used modelling technique is compartmental modelling. Compartment models assume that each tracer molecule is present at each time point in one of a fixed number of compartments. Compartments are defined by anatomical regions (e.g., blood vessels, organs) and specific states of the tracer molecule within one region (e.g., tracer bound to a receptor in the brain). The transition of tracer molecules between the compartments is described by rate constants. A model that fits many brain receptor ligands well is shown in Fig. 7.9a. Transition of the blood-brain barrier (BBB) from arterial blood to brain and backward is described by the rate constants k_1 and k_2 respectively. Tracer movement between the two brain compartments with C_1 (concentration of unbound plus nonspecifically bound tracer) and C_2 (concentration of tracer specifically bound to the target molecule) is described by rate constants k_3 and k_4.

In analogy to the description of electric circuits, the physiological system can be characterized by its transfer function. The TAC of the nonmetabolized tracer in the arterial plasma is considered the input function of the system; the TAC measured by PET in the brain is the response function (Fig. 7.9b). The system itself can be characterized by its response to an impulse function. Accordingly, the response to an arbitrary input function is the convolution of this input function with the response of the system to an impulse function. When modeling the kinetics of a brain PET tracer, the inverse problem must be solved: From the TACs measured in the arterial plasma and in the brain tissue, the underlying rate constants must be estimated for the model chosen. Methods of parameter estimation are described in the literature. Usually,

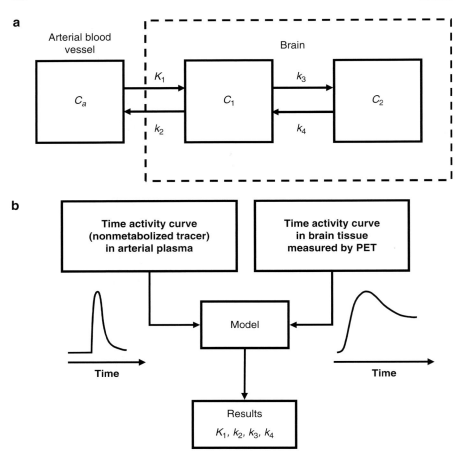

Fig. 7.9 Tracer kinetic modeling with brain positron emission tomographic (PET) data. (**a**) Compartment model for reversibly binding radioligands. Ca is the concentration of the nonmetabolized tracer in the arterial blood plasma. C_1 and C_2 are the tracer concentrations in the compartments 1 (unbound + nonspecifically bound tracer) and 2 (specifically bound tracer) which both belong to the brain. k_1, k_2, k_3, and k_4 are rate constants that define tracer movement between the compartments. (**b**) Determination of rate constants. For linear models, the time activity curve (TAC) in brain (response function) is the convolution of the TAC in arterial plasma (input function) with the impulse response function of the model. In practice, the "inverse problem" must be solved: From measured TACs in the arterial blood (obtained by multiple blood samples) and in brain tissue (obtained by sequential PET acquisitions) and a model configuration such as shown in (**a**), the underlying rate constants are estimated

several models are analyzed for their applicability for a special tracer. Methods to determine which model is most appropriate are also described in the literature [27].

The experimental setup of studies for tracer kinetics modelling is demanding. After tracer injection into a vein, multiple, closely spaced blood samples are taken over an extended time period from an artery. In these samples, the plasma activity concentration is measured. By application of analytical methods, the fraction of plasma activity is determined that is bound to the parent compound (i.e., to the nonmetabolized tracer molecule). The product of these two values defines the input function, that is, the concentration of the radiolabeled, unchanged tracer molecule in arterial plasma. Simultaneously, dynamic brain PET imaging is performed.

The information obtained outweighs the efforts described because modelling improves in a relevant manner the understanding of the relationship between the data acquired by scanning and the underlying physiology and biochemistry. If for a new tracer the parameters defining transition of the BBB and binding are identified, comparison to other known tracers can facilitate the decision to continue or discontinue tracer development. If imaging studies with a new tracer show less-satisfactory results, modelling data may help to identify the special features that need to be improved by modification of the tracer molecule to obtain, for example, altered transition through the BBB or altered binding properties.

If the images made with a new tracer show significant differences between patients and healthy controls, modelling can support the assumption that these differences are in fact caused by differences in the physiology or biochemistry of interest and are not caused by other factors that may be different in the two groups of subjects, such as different arterial input functions due to reduced cardiac function or reduced cerebral blood flow in the patient group.

From algebraic expressions of the rate constants, the parameter distribution of the volume V and the binding potential (BP) can be obtained that are particularly useful for receptor quantification. BP is linked via the radioligand equilibrium dissociation constant K_D to the density of the available receptors B_{avail} in the brain tissues investigated [28].

7.5.4.2 Anatomical Standardization and Voxel-Based Statistical Analysis

The Statistical Parametric Mapping (SPM) software [29] has been designed for the analysis of data coming from various techniques of brain investigation, including SPECT and PET. Images undergo anatomic standardization; that is, the brain images of each subject are transformed into a standard brain ("spatial normalization"). Then, statistical methods are applied on the basis of voxel-wise analysis. The technique is useful for comparison of different groups of persons (e.g., patients vs. healthy volunteers). Regions exhibiting significant group differences can be highlighted in the brain images [30]. With a new tracer, it can be investigated by analyzing small numbers of subjects to what extent the group differences are in agreement with the known pathology of the disease to be diagnosed (e.g., with different receptor densities in specified regions).

Voxel-based methods can also be used to analyze the deviations in a given individual from a group of healthy persons on a statistical basis. This can provide diagnosis in individual persons in a standardized manner. A method developed by Minoshima et al. [31], combining volume image standardization with the projection of cortical activity onto the brain surface (three-dimensional stereotactic surface projection; 3-D-SSP) was shown to be applicable in [^{18}F]-FDG multicenter studies in patients with dementia [32].

7.5.4.3 Visual Analysis of Images

The demonstration of pathologically altered physiology and biochemistry on an image allows a fast diagnosis, which is easy for the imaging expert and is convincing for referring physicians, such as neurologists in the case of brain diseases or surgeons in the case of imaging for cancer metastasis. In the vast majority of applications of current RPs, diagnosis is performed by visual inspection of the images by an experienced nuclear medicine physician. Also for any new tracer, the applicability of performing the diagnosis by visual analysis must be investigated, and rules for image interpretation must be elaborated.

The activity distribution of the amyloid β-targeting tracer BAY94-9172 in the living human brain (Fig. 7.10) is in agreement with the research findings. In the healthy elderly person, activity is seen mainly in the WM, as was the case in autoradiography (Fig. 7.3b). Nonspecific binding to WM also takes place in patients with AD. In the patients with AD, however, there is additional specific binding of the tracer to the gray matter areas that are known to be involved in amyloid β deposition in AD (cerebral cortex and some subcortical areas, such as striatum and thalamus). This specific binding to amyloid β leads to more intense tracer uptake in the gray matter areas than in the WM in patients with AD. By these characteristic image features, patients with AD can easily be distinguished from persons without amyloid β deposits, in whom the images are dominated by nonspecific binding to the WM.

With a new tracer, a normal image database is established, consisting of images performed on healthy persons. Usually, males and females of various age groups are imaged. Further image sets are compiled from the images performed on patients with different diseases or different degrees of severity of a disease. Typical cases and recommendations on how to perform a diagnosis on the basis of the images are provided to the investigators of further clinical studies and to the users of the product after obtaining marketing approval. Special care needs to be bestowed on the description and interpretation of borderline imaging findings.

7.5.4.4 Standardized Uptake Values (SUVs) and SUV Ratios (SUVRs) in PET

To allow comparison between subjects, the activity concentrations measured by PET in one individual person must be adjusted for the total activity administered and for the physical attributes of this subject. A widely used normalization is the SUV, calculated as the ratio of the activity concentration measured in the region of interest (ROI) (kBq/mL) and the injected activity dose (MBq) divided by the body mass (kg). SUVs are successfully used in several clinical applications.

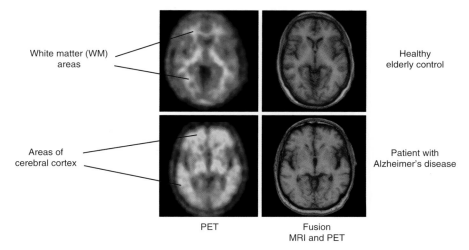

White matter (WM) areas

Healthy elderly control

Areas of cerebral cortex

Patient with Alzheimer's disease

PET   Fusion
MRI and PET

Fig. 7.10 Visual interpretation of images. Imaging with the ^{18}F-labeled tracer BAY94-9172 in an elderly person without signs of dementia (*top*) and in a patient with Alzheimer's disease (AD) (*bottom*). *Left*: Positron emission tomographic (PET) images (transaxial sections) show areas with elevated tracer uptake (*yellow*). For the healthy person, no uptake is seen in the cerebral cortex. Uptake is seen in areas of white matter (WM) containing bundles of nerve fibers. This pattern in healthy persons is typical for current amyloid β PET ligands and does not reflect specific binding to amyloid β. For the patient with AD, a strong uptake is seen in the cerebral cortex (gray matter), which is known to be involved in amyloid β deposition in AD. A color map of the PET image presentation is adjusted for the individual image with *yellow* indicating the highest uptake in each image. *Right*: Fusion of PET images with magnetic resonance (MR) images confirms allocation of tracer uptake to anatomical structures in the brain. (Images courtesy of University Hospital Leipzig, Department of Nuclear Medicine. Director and chairman: Prof. Dr. Osama Sabri, MD, PhD)

However, SUV*s* are still sensitive to biological and technical sources of variability. Since some of these factors add common variance to different regions, a more robust measure is obtained by calculation of SUVRs, that is, ratios between the SUVs of a ROI and a reference region. The presence of similar variation in the ROI and the reference region tends to provide values that are better to compare under different technical settings and in persons with differing physical attributes. For receptor imaging, this approach is meaningful, in particular if there is a reference region that does not express the targeted receptor.

Figure 7.11 shows SUVRs for the amyloid β-targeting tracer BAY94-9172 at several time points up to 3.5 h after injection. The ROI is the frontal cortex, a brain region that is typically involved in the deposition of amyloid β in AD. The reference region is the cerebellum, which is not primarily involved in the disease process. In the patient, SUVRs show steeper initial increase and reach higher values than for the healthy control.

Imaging groups of 15 patients and 15 healthy volunteers showed that SUVRs, calculated from various cortical regions with the cerebellum as the reference region, were significantly higher in patients than in healthy volunteers when determined at

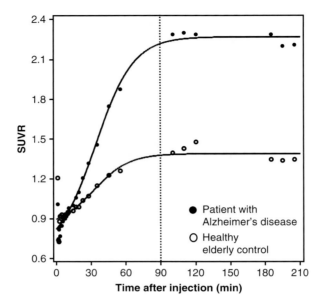

Fig. 7.11 Standardized uptake value ratios (SUVRs): Description of tracer binding in a region involved in the disease by calculation of ratios to a reference region. Imaging with the ^{18}F-labeled tracer BAY94-9172 in a healthy elderly control and in a patient with Alzheimer's disease. Standardized uptake values (SUVs) were determined for the frontal cortex and for the cerebellum as the reference region at various time points after injection. Ordinate: SUVRs, calculated from the SUV of the frontal cortex divided by the SUV of the cerebellum. In the patient, the SUVR shows steeper initial increase and reaches higher values than in the control. The *dotted vertical line* indicates the time after injection when the SUVRs approach stable values. (Images courtesy of Austin Health, Heidelberg, Victoria, Australia. Department of Nuclear Medicine and Centre for PET, C. C. Rowe and V. L. Villemagne)

time points later than about 90 min after injection (dotted vertical line in Fig. 7.11) [33]. SUVR*s* are considered to be suitable parameters for diagnosing individual subjects as either normal in terms of amyloid β load or as having pathological elevation of amyloid β load.

7.5.4.5   Formal Proof of Diagnostic Efficacy

The diagnostic performance of a tracer can be judged from various points of view, including quantitative technical parameters (e.g., signal-to-noise ratios) and subjective assessments such as image quality, easiness of image interpretation, or confidence of the reader regarding the results.

The main aspect in the proof of diagnostic performance, however, is to show the agreement of the diagnostic information obtained with the new tracer with a predefined standard of truth (SoT).

In the ideal case, an absolute SoT (i.e., a diagnostic procedure that has been critically evaluated and documented to identify the true disease state) is available. Usually, histology data obtained from biopsy or surgery specimens are accepted as absolute standards. Often, however, it is not possible to obtain this kind of data in the framework of clinical studies. Biopsies, for example, if they are not needed for the diagnostic workup of the patient, may be regarded as too invasive for the sole purpose of providing data for a clinical study.

In the absence of an absolute standard, a surrogate standard is established, which often is a combination of clinical examinations, laboratory tests, imaging procedures, and others. One important part of the surrogate standard may be the clinical follow-up; that is, the patient is reexamined at appropriate time points until diagnosis can be made with higher certainty (e.g., in the case of progression to a more advanced disease state).

In many cases, a diagnostic question can be defined that can be answered only with "yes" or "no" ("dichotomous" result). The answer given to this question by the new tracer then is compared to the SoT. Questions may refer, for example, to detection (patient has venous thrombosis or not), to characterization (a known breast lesion is malignant or not), or to the spread of a disease (a patient with known cancer has one or more metastases or does not have any metastases).

Per definition, detection of disease is named a "positive" finding, whereas a test result indicating absence of a disease is called "negative." Compared to the SoT, a positive test result can be true (true positive, TP) or false (false positive, FP). Accordingly, a negative test result can be true (true negative, TN) or false (false negative, FN). Table 7.1 shows the four possible combinations of test results and true disease states.

From the frequency distribution of these combinations in a clinical study, several useful parameters can be derived. The two parameters with highest relevance for the characterization of the diagnostic efficacy in the framework of a clinical study are sensitivity and specificity. They are often chosen as the primary variables in clinical studies for diagnostic tests.

Sensitivity is the probability that a test result is positive in subjects who have the disease. Sensitivity is estimated as

$$\text{Sensitivity} = TP/(TP + FN). \qquad (7.6)$$

Table 7.1 Possible combinations of test results and standard of truth (SoT) for a test delivering dichotomous results

		True disease state (SoT)	
		Present	Absent
Result of the new diagnostic procedure	Positive	TP	FP
	Negative	FN	TN

FN false negative, *FP* false positive, *TN* true negative, *TP* true positive.

Table 7.2 Example of study results for a test delivering dichotomous results

		True disease state (standard of truth, SoT)		
		Present	Absent	Sum
Result of the new diagnostic procedure	Positive	TP: 45	FP: 7	52
	Negative	FN: 5	TN: 43	48
	Sum	50	50	100

Sensitivity $= 45/(45 + 5) = 90\%$
Specificity $= 43/(43 + 7) = 86\%$

FN false negative, *FP* false positive, *TN* true negative, *TP* true positive.

Specificity is the probability that a test result is negative in subjects who do not have the disease. Specificity is estimated as

$$\text{Specificity} = TN/(TN + FP) \tag{7.7}$$

Table 7.2 shows an example of test results in a study with 100 subjects.

Other accepted approaches for the proof of efficacy of diagnostic agents (including contrast agents and RPs) (e.g., analyzing the agreement of the results obtained with the new tracer to those obtained with a comparator) are found in health authority guidelines. Such documents also give advice on how to show clinical usefulness beyond the diagnostic performance (i.e., by demonstrating the impact of a new test on diagnostic thinking and on therapeutic decisions) [34]. Guidelines also define requirements regarding randomization (random allocation of patients, e.g., for parallel group comparison of diagnostic tests), blinding (readers of images are blinded for clinical data and other test results), and other methodological questions of clinical studies.

7.5.4.6 Number of Patients in Clinical Studies

The values of sensitivity and specificity displayed in Table 7.2 are *point estimates*, calculated from the numbers of the observed TP, FP, TN and FN cases in a limited number of patients. Such point estimates are influenced not only by the true properties of the diagnostic test but also by chance. The lower the numbers of studied subjects are, the higher is the probability that the influence of pure chance alone leads to large deviations of the point estimates from the "true" values. Like for all biological parameters, true values of efficacy parameters could be obtained in theory by studying an infinitely high number of patients. This is not possible in reality.

However, by calculating sensitivity and specificity as outlined here on the basis of one or several clinical studies, estimates of the true values can be made. A frequently used statistical measure that is provided together with the point estimate is the two-sided 95% confidence interval (CI). For the data shown in Table 7.2 with point estimates for sensitivity and specificity of 90% and 86%, the 95% Clopper–Pearson

CI for sensitivity would be from 78% to 97%, and the 95% Clopper–Pearson-CI for specificity would be from 73% to 94%. If the clinical study could be repeated infinitely often, the true value in 95% of all studies would be in the 95% CI.

With the data presented, it is not allowed to claim sensitivity of 90% and specificity of 86%, as suggested by the point estimates. The adequate claim is that the true values are at least as high as the lower limits of the CIs (i.e., 78% for sensitivity and 73% for specificity). If numbers of patients are increased, the width of the 95% CI becomes smaller, and its lower limit is elevated. This relation between patient numbers and the lower limit of the CI is a reason for relatively high patient numbers in clinical studies.

References

1. Laufer EM, Winkens HM, Corsten MF, Reutelingsperger CP, Narula J, Hofstra L. PET and SPECT imaging of apoptosis in vulnerable atherosclerotic plaques with radiolabeled Annexin A5. Q J Nucl Med Mol Imaging 53, 26–34 (2009).
2. U.S. Department of Health and Human Services, Food and Drug Administration, Center for Drug Evaluation and Research. Guidance for industry, investigators, and reviewers, exploratory IND studies (2006). http://www.fda.gov/downloads/Drugs/Guidance-ComplianceRegulatoryInformation/Guidances/ucm078933.pdf.
3. European Medicines Agency. ICH Topic M 3 (R2). Note for guidance on non-clinical safety studies for the conduct of human clinical trials and marketing authorization for pharmaceuticals. CPMP/ICH/286/95 (2009). http://www.ema.europa.eu/docs/en_GB/document_library/Scientific_guideline/2009/09/WC500002720.pdf.
4. Kirschner AS, Ice RD, Beierwaltes WH. Radiation dosimetry of ^{131}I-19-iodocholesterol. J Nucl Med 14, 713–717 (1973); The authors' reply. J Nucl Med 16, 248–249 (1975).
5. Boxenbaum H. Interspecies scaling, allometry, physiological time, and the ground plan of pharmacokinetics. J Pharmacokinet Biopharm 10, 201–227 (1982).
6. O'Keefe GJ, Saunder TH, Gong SJ, Pathmaraj KU, Tochon-Danguy HJ, Villemagne VV, Krause S, Dyrks T, Dinkelborg L, Holl G, Rowe CC. Comparisons of animal human-translated and human ^{18}F-BAY94–9172 amyloid radiation dosimetry. J Nucl Med 50 (Suppl 2), 1847 (2009).
7. Kolbert KS, Watson T, Matei C, Xu S, Koutcher JA, Sgouros G. Murine S factors for liver, spleen, and kidney. J Nucl Med 44, 784–791 (2003).
8. Franc BL, Acton PD, Mari C, Hasegawa BH. Small-animal SPECT and SPECT/CT: Important tools for preclinical investigation. J Nucl Med 49, 1651–1663 (2008).
9. Liang H, Yang Y, Yang K, Wu Y, Boone JM, Cherry SR. A microPET/CT system for *in vivo* small animal imaging. Phys Med Biol 52, 3881–3894 (2007).
10. Duff K, Suleman F. Transgenic mouse models of Alzheimer's disease: how useful have they been for therapeutic development? Brief Funct Genomics Proteomics 3, 47–59 (2004).
11. European Medicines Agency. ICH Topic E 6 (R1). Guideline for good clinical practice. Step 5. Note for guidance on good clinical practice (CPMP/ICH/135/95) (2002). http://www.ema.europa.eu/docs/en_GB/document_library/Scientific_guideline/2009/09/WC500002874.pdf.
12. Zeng W. Communicating radiation exposure: a simple approach. J Nucl Med Technol 29, 156–158 (2001).
13. Bundesamt für Strahlenschutz [German Federal Authority for Radiation Protection]. Home page http://www.bfs.de.
14. European system for reporting adverse reactions to and defects in radiopharmaceuticals: annual report 2000. Eur J Nucl Med 29, BP13–BP19 (2002).

15. Chianelli M, Mather SJ, Grossmann A, Sobnack R, Fritzberg A, Britton KE, Signore A. ^{99m}Tc-interleukin-2 scintigraphy in normal subjects and in patients with autoimmune diseases: a feasibility study. Eur J Nucl Med Mol Imaging 35, 2286–2293 (2008).
16. Love C, Tronco GG, Palestro CJ. Imaging of infection and inflammation with ^{99m}Tc-fanolesomab. Q J Nucl Med Mol Imaging 50, 113–120 (2006).
17. Loevinger R, Budinger TF, Watson EE. MIRD primer for absorbed dose calculations. Society of Nuclear Medicine, New York. ISBN: 978-0-9320-0438-3 (1988).
18. International Commission on Radiological Protection. Annals of the ICRP. ICRP Publication 60. 1990 Recommendations of the ICRP. Elsevier, New York, ISBN: 978-0-08-041144-6 (1991).
19. International Commission on Radiological Protection. Annals of the ICRP. ICRP Publication 103. Recommendations of the ICRP. Elsevier, New York, ISBN: 978-0-7020-3048-2 (2008).
20. MIRD Pamphlet No. 16: Techniques for quantitative radiopharmaceutical biodistribution data acquisition and analysis for use in human radiation dose estimates. J Nucl Med 40, 37S–61S (1999).
21. He B, Wahl RL, Du Y, Sgouros G, Jacene H, Flinn I, Frey EC. Comparison of residence time estimation methods for radioimmunotherapy dosimetry and treatment planning – Monte Carlo simulation studies. IEEE Trans Med Imaging 27, 521–530 (2008).
22. Sattler B, Barthel H, Seese A, Patt M, Schildan A, Starke A, Reininger C, Rohde B, Holl G, Gertz HJ, Hegerl U, Sabri O. Radiation risk caused by [F18]Bay 94-9172, a new PET tracer for detection of cerebral β-amyloid plaques. J Nucl Med 50 (Supplement 2), 1840 (2009).
23. International Commission on Radiological Protection. Annals of the ICRP. ICRP Publication 23. Reference man: anatomical, physiological and metabolic characteristics. Elsevier, New York, ISBN: 978-0-08-017024-4 (1975).
24. O'Keefe GJ, Saunder TH, Ng S, Ackerman U, Tochon-Danguy HJ, Chan JG, Gong S, Dyrks T, Lindemann S, Holl G, Dinkelborg L, Villemagne V, Rowe CC. Radiation dosimetry of β-amyloid tracers ^{11}C-PiB and ^{18}F-BAY9-9172. J Nucl Med 50, 309–315 (2009).
25. Stabin MG, Sparks RB, Crowe E. OLINDA/EXM: the second-generation personal computer software for internal dose assessment in nuclear medicine. J Nucl Med 46, 1023–1027 (2005).
26. Boroujerdi M. Pharmacokinetics: principles and applications. McGraw-Hill, New York, ISBN: 9780071351645 (2002).
27. Carson EC. Tracer kinetic modeling in PET. In: Valk PE, Bailey DL, Townsend DW, Maisey MN (editors). Positron emission tomography: basic science and clinical practice. Springer, London, 147–179. ISBN: 1 85233 798 2 (2003).
28. Innis RB, Cunningham VJ, Delforge J, Fujita M, Gjedde A, Gunn RN, Holden J, Houle S, Huang SC, Ichise M, Iida H, Ito H, Kimura Y, Koeppe RA, Knudsen GM, Knuuti J, Lammertsma AA, Laruelle M, Logan J, Maguire RP, Mintun MA, Morris ED, Parsey R, Price JC, Slifstein M, Sossi V, Suhara T, Votaw JR, Wong DF, Carson RE. Consensus nomenclature for *in vivo* imaging of reversibly binding radioligands. J Cereb Blood Flow Metab 27, 1533–1539 (2007).
29. Friston KF, Ashburner J, Kiebel SJ, Nichols TE, Penny WD (editors). Statistical parametric mapping: the analysis of functional brain images. Academic Press, London, ISBN: 9780123725608 (2007).
30. Jeong Y, Cho SS, Park JM, Kang SJ, Lee JS, Kang E, Na DL, Kim SE. ^{18}F-FDG PET findings in frontotemporal dementia: an SPM analysis of 29 patients. J Nucl Med 46, 233–239 (2005).
31. Minoshima S, Frey KA, Koeppe RA, Foster NL, Kuhl DE. A diagnostic approach in Alzheimer's disease using three-dimensional stereotactic surface projections of fluorine-18-FDG PET. J Nucl Med 36, 1238–1248 (1995).
32. Mosconi L, Tsui WH, Herholz K, Pupi A, Drezga A, Lucignani G, Reiman EM, Holthoff V, Kalbe E, Sorbi S, Diehl-Schmid J, Perneczky R, Clerici F, Caselli R, Beuthien-Baumann B, Kurz A, Minoshima S, de Leon MJ. Multicenter standardized ^{18}F-FDG PET diagnosis of mild cognitive impairment, Alzheimer's disease, and other dementias. J Nucl Med 49, 390–398 (2008).

33. Rowe CC, Ackermann U, Browne W, Mulligan R, Pike KL, O'Keefe G, Tochon-Danguy
    H, Chan G, Berlangieri SU, Jones G, Dickinson-Rowe KL, Kung HP, Zhang W, Kung MP,
    Skovronsky D, Dyrks T, Holl G, Krause S, Friebe M, Lehmann L, Lindemann S, Dinkel-
    borg LM, Masters CL, Villemagne VL. Imaging of amyloid β in Alzheimer's disease with
    ^{18}F-Bay94-9172, a novel PET tracer: proof of mechanism. Lancet Neurol 7, 129–135 (2008).
34. European Medicines Agency, Committee for proprietary medicinal products (CPMP). Points to
    consider on the evaluation of diagnostic agents, CPMP/EWP/1119/98 (2001). http://www.ema.
    europa.eu/docs/en_GB/document_library/Scientific_guideline/2009/09/WC500003655.pdf.

Part IV
Radiation Detectors for Medical Applications

Chapter 8
Basic Principles of Detection of Ionizing Radiation Used in Medical Imaging

Andrej Studen and Marko Mikuž

8.1 Introduction

The use of ionizing radiation in medical diagnostics became widespread after the dawn of X-ray imaging in the beginning of the past century. The radiation used at present is variable, as are the techniques for its detection. Here, we describe the imaging modalities that use ionizing radiation, basic principles of interaction to which the radiation used is subjected, and statistical and electronic treatment of the measurements.

8.2 Ionizing Radiation Used in Medical Diagnostics

The prevalent diagnostic technique using ionizing radiation is also the oldest: X-ray imaging. Its upgraded variation is computed tomography (CT). This has been complemented by emission imaging: by gamma cameras and single-photon emission computed tomography (SPECT) in the 1950s and by positron emission tomography (PET) in the 1980s. As the most recent development, interoperative beta probes are used for recognition of carcinoma tissues during surgical procedures. Table 8.1 shows relevant information for the types of radiation used in medical diagnostics.

M. Mikuž (✉)
Faculty of Mathematics and Physics, Department of Physics, University of Ljubljana,
1000 Ljubljana, Slovenia
e-mail: Marko.Mikuz@ijs.si

A. Studen
Jožef Stefan Institute, 1000 Ljubljana, Slovenia
e-mail: Andrej.Studen@ijs.si

M.C. Cantone and C. Hoeschen (eds.), *Radiation Physics for Nuclear Medicine*,
DOI 10.1007/978-3-642-11327-7_8, © Springer-Verlag Berlin Heidelberg 2011

Table 8.1 Properties of particles used in medical imaging (nuclear and X-ray)

Source	Photon energy	Half-life	Image modality	Collimation
X-ray tube	10–200 keV	Not applicable	Planar X-ray, CT	Not required (known source position); mechanical (Pb) – grid – for image quality improvement
^{99m}Tc	140 keV	6 h	SPECT	Mechanical (Pb): single/multiple pinhole, slits, coded aperture, ... Electronic: Compton camera
^{111}In	171 and 245 keV	2 days		
^{131}I	364 and 391 keV	8 days		
^{22}Na	Positrons up to 2 MeV 2 photons 511 keV each	2.6 years	PET, SPECT, beta	PET: Collimation on back-to-back photons Beta: No collimation
^{11}C, ^{15}O, ^{18}F		20, 2, 110 min, respectively		

CT computed tomography, *PET* positron emission tomography, *SPECT* single-photon emission computed tomography.

8.2.1 X-Ray Imaging

In X-ray imaging, the object (part of a patient's body) is illuminated with X-ray photons. The X-ray range of the electromagnetic spectrum starts approximately at 1 nm wavelength, corresponding to individual photon energies of 1 keV, and stretches to a few 100 keV. These energies are typical for the atomic properties of matter constituents. The source of X-rays is called an X-ray tube, essentially an electron gun with several tens of kilovolt cathode-to-anode voltage. Figure 8.1 shows a typical spectrum of an X-ray tube; a continuous bremsstrahlung spectrum is superimposed by sharp emission lines characteristic for the anode material. The direction of the photons is selected by a collimator. The imaging strategy is mapping out the absorption coefficient of the imaged object. The detector is placed behind the object and determines the intensity (energy or number) of the photons that pass through the body. The attenuation is proportional to the density and atomic number of tissues traversed (see Sect. 8.3). Conventional X-ray imaging gives a two-dimensional (2D) projection of attenuation on the line along the photon direction. In todays standard CT, both the source and the detector are rotated around a selected axis, and the assembly of 2D images is processed to obtain a three-dimensional (3D) representation of the attenuation in the object.

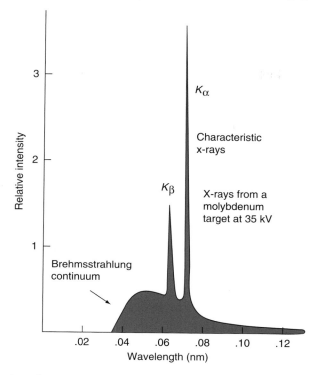

Fig. 8.1 Beam intensity versus wavelength for a molybdenum (Mo) X-ray tube (From [2])

8.2.2 *Emission Imaging*

In emission imaging, the radiation used is gamma rays. These are emitted from nuclear decay of radioactive isotopes. The sources used are called *radiotracers*, specially prepared substances with bound radioactive nuclei. To be used in imaging, the substances should either be or at least mimic normal biomolecules, which circle our bodies. The task of the detector is now to determine the spatial distribution of such molecules in the body. Table 8.1 gives some typical nuclei used in emission imaging classified according to typical imaging modality. Based on the radiation emitted, there are three different imaging modality branches, illustrated in Fig. 8.2:

- In gamma cameras and SPECT, the radioactive source emits one photon at a time. To detect the position of the photon source, a collimator has to be used in front of the sensor. The collimator might be a single pinhole, multiple slits, or multiple pinholes (coded aperture imaging). Recent developments also include electronic collimation based on Compton scattering (Compton camera). The tomographic image is obtained by rotation of the camera around the object.
- In PET, the radioactive source emits positrons (β^+ decay), antiparticles of electrons. The positron annihilates in proximity to its emission, and two 511-keV

Emission imaging

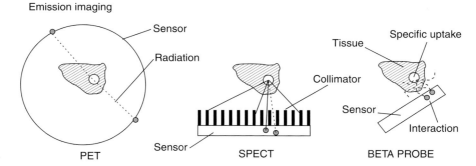

Fig. 8.2 Schematic drawing of most common nuclear emission modalities. *PET* positron emission tomography, *SPECT* single-photon emission computed tomography

photons each are emitted in opposite directions. By detecting the pair of photons in coincidence, a line of possible source positions is obtained.

- In beta probes, we are tracing sources, which emit fast electrons (β^- and β^+ decay). Both rate and energy of the emerging electrons are strongly correlated with the proximity of the detector to the radioactive source.

8.3 Interaction of Photons and Electrons with Matter

There are two types of radiation used for imaging: photons, whether X- or gamma rays, with energies between 1 and up to 511 keV, and electrons. Two types of photon interactions with matter – photoelectric effect and Compton scattering – dominate in the energy range of interest. Both produce fast electrons, not unlike the electrons used in beta probes. We deal with the three effects separately. One common aspect of photon interactions needs spelling out before we go into details. For this, we imagine a flux Φ_0 of photons incident onto a slab of material with thickness d (see Fig. 8.3). Only part of the photons passes through the material unaffected, and their flux Φ_1 reads [1]

$$\frac{\phi_1}{\phi_0} = e^{-\mu d} \qquad (8.1)$$

By this equation, the absorption coefficient μ is defined (with units of 1/m). Since for small μd

$$\frac{\phi_1}{\phi_0} \approx 1 - \mu d \qquad (8.2)$$

the parameter μ roughly measures the fraction of photons that interact per unit thickness of material. As μ tends to be proportionate to the density of the material, it is customary to factor it out by defining the mass absorption coefficient μ/ρ [units $(g/cm^2)^{-1}$] and express the thickness of the material in surface density ρd with units of grams per square centimeter (g/cm^2).

Interaction of photons and electrons with matter

Incoming flux

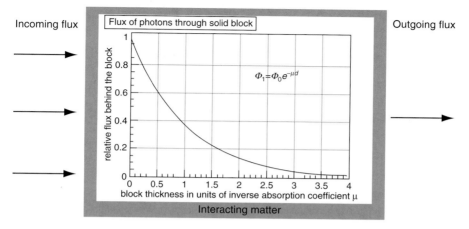

Outgoing flux

Fig. 8.3 Photon attenuation in matter

8.3.1 Photoelectric Effect

8.3.1.1 Properties

The left of Figure 8.4 shows a schematic drawing of the photoelectric effect. The incident photon interacts with both the nucleus and an electron in the orbit of an atom that builds up the interacting medium. The momentum is absorbed by the heavy nucleus, and the energy is transferred to the electron as its kinetic energy. Some energy is required for the electron to escape its bound state in the atom, so the kinetic energy of the interaction electron (also called the photoelectron) is smaller than the energy of the photon. The deposited energy, however, equals that of the incident photon since the vacated state in the atom is gradually filled with electrons in higher energy states. The outcome of the interaction is a fast electron that ionizes matter in the vicinity of the interaction, accompanied with characteristic X-rays from the deexcitation of the atom.

8.3.1.2 Probability

The probability of the photoelectric effect or photoabsorption is shown in Fig. 8.5. The quantity shown on the vertical axis is the attenuation coefficient μ as defined in this chapter. It is common to express the probability in terms of the scattering cross section σ (measured in square meters) per atom of interacting matter. The relation between the two is

$$\mu = \frac{\rho N_A}{A}\sigma \tag{8.3}$$

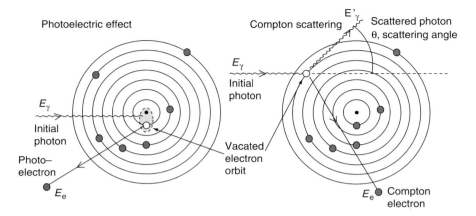

Fig. 8.4 Most common photon interactions in matter for photons used in medical imaging

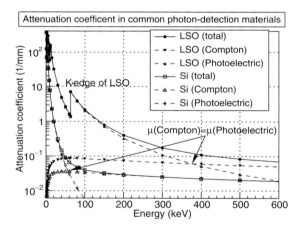

Fig. 8.5 Attenuation coefficient in silicon (semiconductor) and lutetium orthosilicate (LSO; scintillator) in the energy range of photons used in medical imaging. Attenuation is split into photoabsorption (PA), Compton scattering (CS), and total (T). Also marked is the point of equal probability of photoabsorption and Compton scattering (Data from [3])

where ρ is the material density, N_A is Avogadro's number ($N_A = 6.023 \times 10^{26}$), and A is the atomic mass of the atom where the interaction occurred. As mentioned, μ is proportional to material density for similar atomic mixtures.

Observing Fig. 8.5, we can see that there is a sharp rise in photoelectric effect probability when the energy of the incident photon exceeds the K line of the atomic spectra. The photoelectric effect relies on the nucleus to absorb the momentum of the incident photon. The electrons in the nearest K shell of the atom have the best "view" of the nucleus and hence are the most susceptible to the photoelectric effect. The result is a jump by an order of magnitude in scattering probability when photons can penetrate so close to the nucleus of the atom. For energies above the "K edge", the interaction probability is roughly proportional to [1]

$$\sigma \propto \frac{Z^{4-5}}{E^{3.5}} \tag{8.4}$$

where Z is the atomic number of the interacting atom, and E is the energy of the photon. The exponent in Z^N varies between $N = 4$ and $N = 5$ in the (photon) energy range of interest.

8.3.2   Compton Scattering

8.3.2.1   Properties

Description of Compton scattering followed on the success of the quantum mechanical description of the photoelectric effect. Here, we treat both the bound electron and the photon as quasi-free particles with energy and momentum and observe their elastic collision as shown in Fig. 8.4 on the right. Both (relativistic) energy and momentum are conserved, giving rise to the famous Compton equation

$$\sin^2(\theta/2) = \frac{1}{2\gamma} \frac{\varepsilon}{1 - \varepsilon} \tag{8.5}$$

where θ is the scattering angle of the photon, $\gamma = E_\gamma/m_e c^2$ is the energy E_γ of the initial photon scaled to the electron rest mass ($m_e c^2 = 511\,\text{keV}$), and ε is the proportion of energy E_e transferred to the Compton electron $\varepsilon = E_e/E_\gamma$. As opposed to photoabsorption, the energy deposit depends on the scattering angle, giving rise to a continuous spectrum of energy loss, as seen in Fig. 8.9. Relativistic kinematics limits the maximum energy transfer to

$$\varepsilon_{\text{max}} = \frac{2\gamma}{1 + 2\gamma} \tag{8.6}$$

yielding 49.8 keV as the Compton edge of the $^{99\text{m}}$Tc source; its spectrum is portrayed in Fig. 8.9.

8.3.2.2   Probability of Interaction

Since the nucleus is not involved in the interaction, the probability is proportional to the number of electrons Z per atom. The energy dependence is weak in the region of interest, making Compton scattering the dominant interaction for larger photon

Table 8.2 Photon energies for which photoabsorption and Compton scattering are equally probable

Material (acronym or chemical formula)	Energy at which Compton scattering and photoabsorption are equally probable (keV)
Bismuth germanate oxide (BGO)	450
Lutetium/lutetium yttrium orthosilicate (LSO/LYSO)	400/300
Sodium iodide (NaI)	250
Crystalline silicon (Si)	60

energies (see Fig. 8.5). Since the photoelectric effect drops as $1/E^{3.5}$, the dominance is quickly asserted beyond the equal probability point, the energy of which depends heavily on Z, the atomic number of the scattering material. See Table 8.2 for typical values of this energy for some detector materials.

8.3.3 Interaction of Fast Electrons and Secondary Ionization

Whether the original particle is either an electron or a photon that underwent one of the photon interactions described in detail in the preceding section, both result in an electron energetic enough to ionize the matter it traverses. The ionization continues until all kinetic energy of the original particle is dissipated and transferred to a number of detectable particles, their nature depending on the type of detector used.

8.3.3.1 Scintillator

In scintillators, the ionization is converted to photons of visual light. The number of photons created is proportional to the energy loss of the fast electron, with typical numbers of 8 (bismuth germanate oxide, BGO), 25 (lutetium orthosilicate, LSO), and 38 (NaI) photons per kilo electron volt. A photoabsorbed annihilation photon will thus yield 4,000, 12,500, and 20,000 visual photons, respectively.

8.3.3.2 Semiconductors

In semiconductors, the ionization is converted to electron–hole pairs. Their count is proportional to energy loss, with the proportionality factor given as the mean energy needed for pair creation with values of 2.96 eV (Ge), 3.6 eV (Si), and 5 eV CdZnTe. A photoabsorbed annihilation photon will thus create 173,000, 141,000 and 102,000 electron–hole pairs, respectively.

8.3.3.3   Photographic Emulsion

The secondary ionization is spent to break the silver–halide bond, forming clusters of neutral atomic silver within the emulsion. Here, the dependence of the number of silver atoms on the energy of the particle is in general nonlinear and becomes saturated quickly.

8.3.3.4   Range

The ionization of the fast electron is not instantaneous. Rather, the fast electron travels a distance in matter, denoted as the *range*, ionizing along the whole length of the track until all of its kinetic energy is lost. Whereas in X-ray applications the range can be safely ignored, it stands as an inherent spatial resolution limit in emission imaging. For example, in silicon used as a PET detector, the electrons can travel up to 300 μm from the photon interaction point, as for 340-keV electrons at the Compton edge of annihilation photon interaction. The same applies to the emitted positron range in tissue, which can be on the order of several hundred micrometers.

8.4   Statistical Treatment of Measurements

Imaging with quantum particles, like photons, will be random in nature. This means, for example, that the number of detected photons will vary for repetitive measurements. The only verification of the measurement result can be performed through statistical tests. Let us take a brief look at effects of the underlying quantum nature on the photon detection process.

Think of a setup like that given in Fig. 8.3. Not only is the flux equation (Eq. 8.1) valid for the number of photons exiting the material, but also it is valid for any single photon. Therefore, we can think of the exponent as q, the probability that the photon will pass through the slab, and equivalently, $p = 1 - q$ is the probability that the photon will interact in matter. For N photons incident on the material N_1, the number of photons that interacted, is distributed according to the binomial distribution

$$P(N_1; p, N) = \frac{N_1!}{(N - N_1)!N_1!} p^{N_1} (1 - p)^{N - N_1} \tag{8.7}$$

exhibiting the expected average $<N_1> = pN$. If the number of incident photons is large ($N \to \infty$) and the probability p of the photon interacting is small ($p \to 0$, say for thin samples), while the product $\eta = pN$ representing the mean value of interacting photons ($<N_1>$) remains finite, then the binomial distribution turns into the Poissonian distribution:

$$P(N_1; \eta) = \frac{\eta^{N_1}}{N_1!} e^{-\eta} \tag{8.8}$$

For sufficiently large mean values η, which in practical terms happens already at η ~ 10, the distribution turns into the Gaussian probability function $g(x; x_0, \sigma)$, with both x_0 and σ^2 equal to η. In fact, due to the central limit theorem, any measurement that is comprised of a large number of steps with different probability distributions will in the limit follow the Gaussian distribution:

$$g(x; x_0, \sigma) = \frac{1}{\sigma \sqrt{2\pi}} \exp \left[-\frac{(x - x_0)^2}{2\sigma^2} \right] \tag{8.9}$$

The Gaussian distribution is a probability distribution function, defined for a continuous variable x, even when the underlying variable is in fact an integer, such as for photon counting. Then, the approximate probability for the given integer value is calculated as the integral over an interval with a size of 1 centered on the sought integer. The parameters – average x_0 and variance σ^2 – can be estimated from N measurements using standard estimation techniques:

$$x_0 = \frac{1}{N} \sum_{i=1}^{N} x_i; \quad \sigma^2 = \frac{1}{N-1} \sum_{i=1}^{N} (x_i - x_0)^2 \tag{8.10}$$

or determined by a fit to the measured histogram.

Once the measured distribution with its parameters is known, we can compare it to a hypothesis. This process is equivalent to the judgment whether a gray spot in a black region of the X-ray image is due to a random excursion of background ("normal" in Fig. 8.6) or due to a significant tissue modification ("abnormal"). This rudimentary "by eye" comparison can be performed more rigorously using statistical tests, for example, the χ^2 test. Once a particular χ^2 value is determined, the statisti-

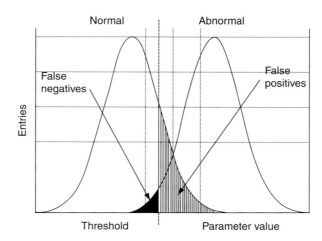

Fig. 8.6 False-negative/false-positive rates for a cut with specified statistical significance applied to a distribution according to the relevant parameter

cal significance of the gray spot can be calculated or read from tables. The threshold set on the level of significance leads to the ultimate statistical dilemma (illustrated in Fig. 8.6), termed in many ways: ratio of false positives versus false negatives, specificity versus sensitivity, or receiver-operating-characteristic (ROC) curve.

8.5 Detector Assemblies

The final detection units require additional components on top of the sensing material, most notably electronics and mechanics. The latter includes collimators and mechanical gantries, which are as a rule specific to a detection task and we do not delve into those. This section deals with electronics and its impact on detector performance.

As we have seen, the result of interaction in the sensor will be a number of particles (visual photons, electron–hole pairs) that have to be counted. There are two approaches: *current* mode, in which one counts for a fixed exposure time, ignoring individual events, and *single-photon counting* mode, in which events are treated in an individual manner (e.g., storing time and energy of interaction for each photon impact). The current mode is simpler to implement and often meets the requirements, for example, in X-ray and CT, for which the source provides plenty of photons and this large statistic diminishes counting fluctuations. A typical current mode device is photographic emulsion (film). Once exposed to radiation, the film is developed using standard photographic techniques, and no additional processing is required. The same principle is valid for phosphorus plates or flat-panel detectors. Further information on those is to be found in Chap. 10 on new detectors. The discussion next deals with electronic processing in single-photon counting mode.

8.5.1 Signal Processing

The single-photon counting mode is more complex since readout has to be performed for each event. However, it provides the most detailed information, useful especially when count statistics are low, which is typical for emission imaging. The task of the readout is to preserve the "fragile" information contained in raw signals, to maximize efficiency, and to reduce dead time, thus quickly managing collected data (transfer logistics, storage). Here, we are talking about fast, low-noise signal-processing techniques.

A typical readout chain, portrayed in Fig. 8.7, involves a charge-sensitive preamplifier, which converts charge pulses into voltage steps, and a shaper, which is essentially a bandwidth-limiting filter. Looking more closely, Fig. 8.8 shows the composition of a typical *CR-RC* shaper, the name coined as a string of abbreviated circuit names that are used. The output of such a shaper is an electronic pulse whose amplitude (in volts) is proportional to the number of particles (electrons, photons)

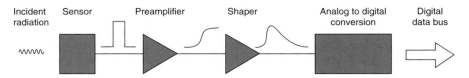

Fig. 8.7 Schematic drawing of the generic readout chain for single-photon readout. (After [4])

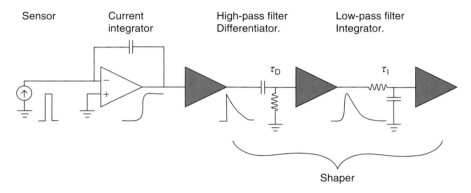

Fig. 8.8 Detailed schematics of the first part of the readout chain. (After [4])

created by the incident radiation. Even for a monoenergetic radiation source, the output amplitude will vary approximately according to the Gaussian distribution; the noise figure is equal to the parameter σ quantifying the precision of the measurement. Rather than giving the value in millivolts (which often has little or no meaning), it is instructive to express it as the equivalent noise charge (ENC) (the number of raw carriers at input giving a signal equal to the noise) or as energy resolution, which is the full width at half maximum (FWHM) of the distribution of readings for a fixed energy of the incident radiation.

The total noise σ_t is composed of two parts, which are not correlated and thus add in quadrature:

$$\sigma_t^2 = \sigma_S^2 + \sigma_E^2 \tag{8.11}$$

where σ_S is the statistical counting noise, and σ_E is the variation of the baseline attributed to the electronics. The statistical noise is related to fluctuations in the number of particles N created during secondary ionization and presents the inherent limit to energy resolution. One would expect the statistical noise in ENC to follow Poisson statistics with $\sigma^2 = N$. However, because the individual carriers in N are constrained by the total energy loss, the statistical noise is scaled with the Fano factor F:

$$\sigma_S^2 = FN \tag{8.12}$$

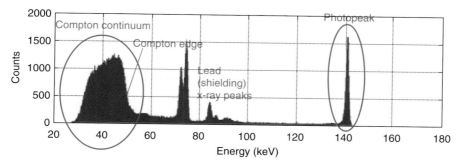

Fig. 8.9 Spectrum of ^{99m}Tc as recorded by a silicon detector. The full width at half maximum (FWHM) resolution of the 140.5-keV photo peak was 1.4 keV, corresponding to an equivalent noise charge (ENC) of 165 e (From [5])

where the value of F depends on the detailed process of secondary ionization. In silicon $F = 0.1$, while for most scintillators, $F \sim 1$.

The electronic noise can be parameterized as follows:

$$\sigma_E^2 = A \cdot C_{total} \frac{1}{\tau} + B \cdot I_{dark}\tau + D \tag{8.13}$$

where A, B, and D are constants related to specific amplifiers; C_{total} is the total capacitance seen by the preamplifier; I_{dark} is the current flowing into the preamplifier when no signal is present. Here, τ is the shaping time; optimum filtering is achieved with equal differentiator (τ_D) and integrator (τ_I) characteristic times in the CR-RC circuit, and their value is then denoted as shaping time. Taking this expression as a function of the shaping time, τ of the system can be optimized for minimum noise.

As a numerical example, let us analyze the spectrum of ^{99m}Tc in Fig. 8.9, recorded by a silicon detector. ^{99m}Tc emits photons with 140.5-keV energy. If the photon interacts through photoabsorption, 38,000 electron–hole pairs are created. Assuming Possonian statistics, the best possible resolution would be $\sqrt{(38,000)} \sim 200$ electrons ENC. However, the measurement exhibits an ENC of 165 electrons only. In fact, the Fano factor must be taken into account; thus, the resolution is composed of 62 electrons statistically and 152 electrons of the prevailing electronic noise. A shaping time of 3 μs was used.

8.5.2 Dead Time

Dead time T is the time when the sensor is unable to collect data. It is usually dominated by the time required to process data and relay it to downstream electronics. For example, we should always treat the shaping time as dead time since further events

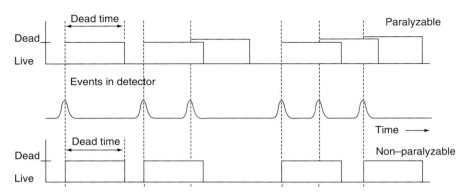

Fig. 8.10 Signals in a paralyzable and nonparalyzable detector system

during that time will be included in the preceding event. The desire for high-rate operation might therefore compromise the optimum noise setting.

There are two extreme models of sensor behavior in terms of dead time illustrated in Fig. 8.10: The *paralyzable* model assumes that events appearing during dead time provoke additional dead time, effectively paralyzing the readout; the *nonparalyzable* model assumes that events during dead time are simply ignored. Most detector systems actually perform between the two extremes. The relation between the true (n) and observed (m) rate is given by

$$n = \frac{m}{1 - mT} \quad \text{for the } \textit{non-paralyzable} \text{ model, and} \tag{8.14}$$

$$m = n \cdot e^{-nT} \quad \text{for the } \textit{paralyzable}. \tag{8.15}$$

Note that in the paralyzable model the equation can only be solved iteratively and yields two ambiguous solutions for the true rate n at each observed rate m.

8.6 Example: Scatterer for a Compton Camera

8.6.1 The Compton Camera Principle

The Compton camera is regarded as an alluring conceptual alternative to the conventional gamma camera. The idea itself is far from recent [6]; however, the severe technological requirements behind the relatively simple concept blocked developments of devices for medical imaging at the level of preclinical prototypes [7–9]. The concept was, however, successfully implemented in astronomy [10] and homeland security [11].

The concept is illustrated in Fig. 8.11. The lead collimator of a standard gamma camera is replaced with an electronic collimator. The electronic collimator is an

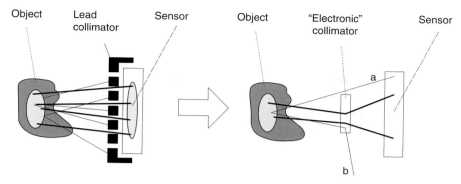

Fig. 8.11 The Compton camera principle. The standard gamma camera with lead collimator (*left*) is replaced with an electronic collimator (*right*)

additional piece of sensitive material. The emitted gamma ray is required first to interact in the collimator/scatterer, undergoing Compton scattering, and the scattered photon must interact in the second detector. The second detector could be just the detector of the conventional gamma camera with its collimator removed. In the picture, the gamma rays that contribute to the recorded image are thickened. Obviously, the standard gamma camera ignores rays that interact in the collimator; conversely, the Compton camera requires them to interact in the collimator/scatterer first. Note in the figure that the rays that interact in the setup but are ignored either interact (a) only in the second detector or (b) only in the scatterer. When both interactions are recorded, the direction of the scattered gamma ray can be reconstructed from both impact positions. An additional piece of information is required – either the energy of the electron excited by the Compton interaction or the energy of the scattered photon. The energy resolution of scintillators used in gamma cameras is usually too poor for this purpose; therefore, we will assume that the electron energy is measured in the scatterer. With known energy of the scattered electron and the Compton equation (Eq. 8.5), the position of the radioactive source can be limited to a cone with its apex at the interaction point in the scatterer and the opening angle equal to the scattering angle in Eq. 8.5. Intersection of multiple cones yields the distribution of sources within the object, as illustrated in Fig. 8.12 for a point source.

The rationale behind the principle deals with limitations of the lead collimator. As higher resolution requires narrower collimators, which subtend a more limited solid angle, there is an inherent trade-off between resolution and efficiency. At low energies of the incident radiation, a high-resolution collimator absorbs most of the photons, thus reducing information content per radioactive dose absorbed in the patient. At higher energies, it performs even worse as more of the photons penetrate the collimator, adding artifacts to collected images. Another aspect is the relative bulkiness and weight of the collimator-equipped setup, preventing its use in space-limited areas. The simulations of the Compton camera have shown great potential improvements in all areas mentioned, and their performance, contrary to collimators, even improves with increasing gamma ray energy.

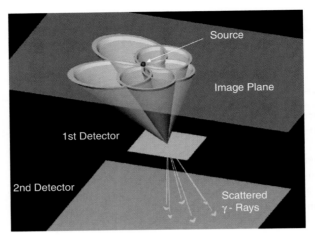

Fig. 8.12 Point source reconstruction from multiple events recorded in a Compton camera

8.6.2 Considerations Regarding Silicon as an Example for the Sensitive Material for a Compton Camera Scatterer

This section shows how principles of ionizing radiation detection (used in medical imaging, as per the title of the chapter) can be used in design and implementation of a detector for medical imaging. The section deals with individual requirements dictated by the principles.

8.6.2.1 Spatial Resolution

The track of the scattered photon will be determined by the resolution of the interaction points in both sensors. Ignoring the resolution of the second detector (assuming a standard gamma camera detector), the scatterer should be able to provide excellent impact resolution properties. There are multiple detectors capable of resolving position, but none is better than solid-state detectors. In fact, it was silicon in which track resolution down to 1 μm was demonstrated, using suitable segmentation of readout electrodes. For the Compton camera, the bar need not to be set so high, but resolutions in the range of 1–2 mm, comparable to standard gamma cameras with a high-resolution collimator, are trivially achievable.

8.6.2.2 Energy Resolution

The Compton equation (Eq. 8.5) implies that the precision of measured electron energy will be migrated into the scattering angle and finally source position resolution. The relation is

$$\sigma_\theta = \frac{(1 + \gamma(1 - \cos\theta))^2}{E_\gamma^2 m_e c^2 \sin\theta}\sigma_{Ee}.$$ (8.16)

There are three pieces of information to be noted in this equation:

1. The energy resolution is of crucial importance. Again, this points to semi-conductor detectors, for which germanium would be best, but silicon is close behind.
2. The influence of limited energy resolution σ_{Ee} on scattering angle resolution σ_θ drops quadratically with energy of the initial photon E_γ, which implies better performance at high-impact photon energy, in strong contrast to increasingly worse performance (penetration) of lead collimated systems.
3. The relation has a characteristic $1/\sin\theta$ shape, implying bad performance for both small and large scattering angles and providing an optimum imaging window for scattering angles around $90°$.

8.6.2.3   Ratio of Compton Interactions

All interactions in the scatterer that are not Compton scattering do not contribute to the final image. From discussion presented in Sect. 8.3.1 and 8.3.2, the ratio of Compton scattering and photoelectric effect will drop with energy as about $1/E^{3.5}$ and approximately as $1/Z^{3-4}$ with respect to the atomic number. Clearly, low-Z elements (silicon) and higher energies are favorable.

8.6.2.4   Doppler Broadening

The validity of the Compton relation (Eq. 8.5) is limited by Doppler broadening. The term was coined because of the close analogy of classical sound distortion in flowing media. There is an alteration of scattered photon energy (or frequency if wave analogy is used) from the value inferred from Eq. 8.5 since the interaction electron is moving within the atom. The electron motion is random, and the net effect is a broadening of the scattering angle around the central value determined by Eq. 8.5. The broadening is related to the electron distributions in the target atom, with approximate proportionality to atomic number and the number of valence electrons. It turns out [12] that silicon has very small Doppler broadening, much narrower than germanium or other high-Z materials.

8.6.2.5   Other Characteristics of Silicon

On the benefits side for silicon, one should note excellent processing and design techniques and reasonable cost, all fueled by developments in chip and processor fabrication. It is robust and sturdy enough to be handled in sensitive environments (hospitals). Concerning the drawbacks, the low atomic number of silicon, which is beneficial for the high Compton ratio and low Doppler profile, implies a small attenuation coefficient of $\mu \approx 0.02/\text{mm}$ for a 511-keV annihilation photon. For decent efficiency, several millimeters of sensitive silicon are required. Due to technology limitations, the thickness of a single sensor is limited to 1–2 mm, and a stack of several detectors needs to be employed, increasing the count (and cost) of required readout electronics and support structures.

8.6.3 Scatterer Characterization

Based on these theoretical considerations, a scatterer prototype has been constructed [13, 14]. The sensor was processed on a 1-mm thick silicon wafer. The silicon was segmented into pads with a size of 1.4×1.4 mm, each pad effectively functioning as an independent p^+-n diode. The pad size was chosen as a compromise between the required spatial resolution (1–2 mm) and the number of required readout channels per detector area. An additional metal layer is put on top of the diode structure, routing the signals from pads to the bonding connection pad at the detector side. The diodes are reverse biased, and the sensor depletes of thermally excited carriers at 150 V. The current flowing through each pad at those voltages is approximately 100–200 pA, the range covered by variation across the detector surface (Fig. 8.13).

An application-specific integrated circuit (ASIC) is used to process the signals. In this case, an off-the-shelf ASIC VATAGP3 by GammaMedica-Ideas was used.

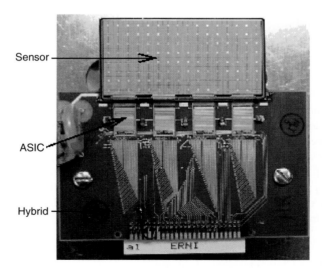

Fig. 8.13 Photograph of a silicon module used for a Compton camera scatterer. The building blocks (512 pad sensor, VATAGP3 ASIC [application-specific integrated circuit], custom made hybrid) are marked with *arrows*. On the sensor, a pattern of metal traces, connecting sensor pads to connection pads at the edge, can be observed

Each VATAGP3 circuit hosts 128 parallel electronic channels, each equipped by a preamplifer feeding in parallel a pair of CR-RC shapers:

- The *fast shaper* used to derive a trigger pulse. Its shaping time is kept short (150–200 ns, adjustable) to optimize timing, and its output is fed to a leading-edge discriminator. A common coarse threshold is applied to the chip, and each channel has a 3-bit DAC (digital-to-analog converter) to fine adjust its threshold to compensate for the gain and baseline offset.
- The *slow shaper* has a shaping time optimized for lowest possible noise (minimum of Eq. 8.13) for energy resolution. At the final setting of 3 μs, noise due to capacitance was still dominating. A large noise variation was observed following the length pattern of on-detector trace lengths (capacitance), connecting pads, and readout electronics.

The scatterer was characterized by measuring spectra of calibration sources (^{241}Am, ^{99m}Tc). The most challenging part proved to be the energy resolution; the final performance could only be achieved after a complete redesign of the support electronics on the hybrid. At the energy resolution of 1.4 keV FWHM as demonstrated in Fig. 8.8, the overall performance in terms of noise resolution trade-off was comparable to collimated systems for 140.5-keV ^{99m}Tc photons already for a single 1 mm thick detector plane. The development of multiple plane detectors is still in progress.

References

1. G. F. Knoll. Radiation Detector and Measurement. 3rd ed. Wiley, New York. 1999.
2. HyperPhysics project, http://hyperphysics.phy-astr.gsu.edu/Hbase/hframe.html.
3. M. J. Berger et al. XCOM: Photon Cross Sections Database. NIST Standard Reference Database 8 (XGAM). http://physics.nist.gov/PhysRefData/Xcom/Text/XCOM.html.
4. H. Spieler. Semiconductor Detector Systems. Oxford University Press, New York. 2005.
5. S. J. Park et al. A prototype of very high-resolution small animal PET scanner using silicon pad detectors. Nucl Inst Meth A 570(3), 543–555, 2007.
6. R. W. Todd et al. A proposed γ camera. Nature 251, 132–134, 1974.
7. J. W. LeBlanc. C-SPRINT: a prototype Compton camera system for low energy gamma ray imaging, IEEE Trans Nucl Sci 45, 943–949 (1998).
8. S. Kabuki et al. Development of electron tracking Compton camera using micro pixel gas chamber for medical imaging. Nucl Inst Meth A 580 (2), 1031–1035, 2007.
9. H. Seo et al. Feasibility study on hybrid medical imaging device based on Compton imaging and magnetic resonance imaging. Appl Radiat Isot 67 (7–8), 1412–1415, 2009.
10. P. F. Bloser et al. The MEGA project: science goals and hardware development. New Astron Rev 50 (7–8), 619–623, 2006.
11. L. Mihailescu et al. SPEIR: A Ge Compton camera. Nucl Inst Meth A 570 (1), 89–100, 2007.
12. A. Zoglauer et al. Doppler broadening as a lower limit to the angular resolution of next generation Compton telescopes. Proc SPIE 4851, 1209–1220, 2003.
13. A. Studen et al. Development of silicon pad detectors and readout electronics for a Compton camera. 501 (1), 273–279, 2003.
14. E. Cochran et al. Performance of electronically collimated SPECT imaging system in the energy range from 140 keV to 511 keV. Nuclear Science Symposium Conference Record, 2008. NSS '08. IEEE, 4618-4621, 2008.

Chapter 9
Scintillators and Semiconductor Detectors

Ivan Veronese

9.1 Scintillators

9.1.1 Basis of Detection and Requirements

Various processes occur during the detection of ionizing radiation within a scintillator, and proper detection designs are needed [1–3]. As a consequence of the interaction of radiation with the scintillation material, ionisation and excitation processes arise, and the energy (or part of it) of the incoming radiation is transferred to the atoms and molecules of the scintillator. Following deexcitation processes, photons originate in the ultraviolet/visible (UV/VIS) region of the electromagnetic spectrum, light that must be collected and converted in a suitable electric signal. In many cases, light collection simply may be obtained by coupling the scintillator directly with an optical detector, typically a photomultiplier tube (PMT). In other cases, depending on the particular application or measurement geometry, a light guide is required, which efficiently transmits the light emitted by the scintillator to the optical device. Finally, light photons are converted into electrons, and the resulting basic electric signal is amplified and properly processed. Let us consider in more detail the scintillation conversion mechanism in a wide band-gap material. This process may be explained by considering the energy band structure of an activated crystalline scintillator. An inorganic scintillator is indeed usually a crystalline solid containing a small amount of dopant, acting as a luminescent centre, which creates energy levels within the forbidden band between the valence band and the conduction band. Moreover, the natural impurities and defects present in the crystal are the origination of other energy levels, which may act as traps during the charge transport.

I. Veronese (✉)
Dipartimento di Fisica, Università degli Studi di Milano and INFN, Sezione di Milano,
Via Celoria 16, 20133 Milan, Italy
e-mail: ivan.veronese@unimi.it

M.C. Cantone and C. Hoeschen (eds.), *Radiation Physics for Nuclear Medicine*,
DOI 10.1007/978-3-642-11327-7_9, © Springer-Verlag Berlin Heidelberg 2011

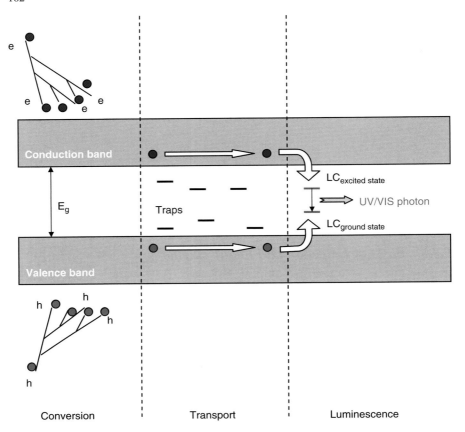

Fig. 9.1 Scintillation mechanism in a wide band-gap material. *UV/VIS* ultraviolet visible

The transformation of the ionising radiation to UV/VIS photons is a complex process that may be divided into three consecutive stages: conversion, transport, and luminescence. During the conversion process, the interaction of the primary ionising radiation (e.g., high-energy photons) with the lattice of the scintillator material creates many electron–hole pairs that thermalise in the conduction and valence band, respectively. In the following step, electrons and holes migrate through the material. During this transport, repeated trapping at defects may occur, and energy losses are probable due to nonradiative recombination processes. Finally, the last stage, luminescence, consists of consecutive radiative recombination processes of the electrons and holes at the luminescent centres, which in general are produced in an excited state. Following deexcitation process, UV/VIS photons are originated. A sketch of the scintillation mechanism is shown in Fig. 9.1.

On the basis of the processes involved in the detection of ionizing radiation with scintillation detectors, it is possible to delineate the main properties that characterise a good scintillator [1–4]: light yield, linearity, transparency, decay time, emission spectrum, chemical stability and radiation hardness, density and effective atomic number.

9.1.1.1 Light Yield

The conversion of the absorbed energy from the incoming ionizing radiation into UV/VIS photons should occur with high scintillation efficiency. In general, the number of scintillation photons N_{ph} emitted when an amount of energy E is absorbed in a scintillator due to interaction with a high-energy photon (X- or gamma ray) or other ionizing radiation is given by

$$N_{ph} = N_{e-h} \cdot S \cdot Q = \frac{E}{\beta \cdot E_{gap}} \cdot S \cdot Q$$

The term $\frac{E}{\beta \cdot E_{gap}}$ represents the number of thermalised electron–hole pairs N_{e-h} produced in the interaction process; E_{gap} is the band-gap energy between valence and conduction band; and β is a parameter introduced to take into account the fact that the average energy required to produce one thermalised electron-hole pair (Ee-h) is higher than the band-gap energy. Typically,

$$E_{e-h} = \beta \cdot E_{gap} \cong 2 - 3 \cdot E_{gap}$$

The term S is the transport efficiency of the electron–hole pairs to the luminescent centre (or, similarly, the transfer efficiency of the absorbed energy to the luminescent centre). It is affected by nonradiative processes, which are generally dependent on the material, temperature, and any imperfections or impurities that may be present in the lattice structure. The term Q is the quantum efficiency of the luminescent centre, that is, the efficiency for photon emission when the luminescent centre is excited.

9.1.1.2 Linearity

The conversion of the absorbed energy into light should be linear over a wide range. The proportionality between the emitted light and the absorbed energy can be affected by saturation phenomena occurring at high energy levels. In the case of X- or gamma ray detection, the energy dependence of the light yield is partly due to rapid changes of the attenuation coefficient around the K and L edges of the elements constituting the scintillator, but it also follows from the nonequal conversion efficiency of the photoelectric and Compton scattering effects, which becomes progressively more important with increasing energy of the incoming ionising radiation [1]. Effects of nonproportionality arise also in case of particle detection because of the dependence of the scintillation efficiency on the linear energy transfer (LET).

An important consequence of the loss of linearity is that the energy resolution of a scintillator material is noticeably degraded with respect to the intrinsic limits based purely on statistical considerations.

9.1.1.3 Transparency

The scintillator should be transparent to the emitted photons, and effects of self-absorption should be minimised. This property is, in general, obtained by the presence of the activators in the lattice structure since the direct deexcitation of electrons from the conduction band to the valence band would cause the origination of energetic photons (i.e., not in the UV/VIS region) with a high probability of self-absorption.

9.1.1.4 Decay Time

A good scintillator should be fast; it should therefore have a short decay time. The kinetics of the light response of a scintillator depends on the transport and luminescence stages of the scintillation conversion mechanism. These two processes are indeed much slower with respect to the initial conversion, which is concluded within less than 1 ps.

In the simplest case of monoexponential decay, the emission intensity $I(t)$ is given by

$$I(t) = I_0 \cdot e^{-t/\tau}$$

where I_0 is the initial intensity, and τ is the decay time. This parameter strongly depends on the activator used to dope the scintillation crystal. While for the parity or spin forbidden transitions for the most rare earth ions decay times are typically on the order of several tens of microseconds up to milliseconds, in the case of the allowed 5d–4f transition of Ce^{3+} and Pr^{3+}, the value scales down to tens of nanoseconds, and fully allowed singlet–singlet transitions in organic molecules are still about ten times faster [1].

Due to the mentioned retrapping processes occurring during the transport stage of the conversion mechanism, the time dependence of luminescence is often further complicated by slower nonexponential components. It is important to note that the presence of afterglow in a scintillator may affect not only the timing performance of the detector but also may make the signal-to-background ratio worse.

9.1.1.5 Emission Spectrum

A good scintillator should be characterised by an efficient spectral matching between its emission spectrum and the optical detector sensitivity curve. The position of the emission spectrum is important for the choice of the photodetector; according to classical criteria, the near-UV blue emission is suitable for a PMT, while for a photodiode the green-red spectral region was considered the best. Recently, however, UV-sensitive photodetectors based on SiC, GaAlN, and diamond were developed, offering a wide choice of detection configurations [1].

9.1.1.6   Chemical Stability and Radiation Hardness

A good scintillator should have good chemical stability and radiation resistance. Chemical stability concerns mainly the hygroscopicity of some crystals (like NaI:Tl, CsI:Na, LaBr$_3$:Ce), which may limit their operation for a long time in air. However, today sealing methods are well developed; therefore, chemical reactions due to the attraction of water molecules can be easily avoided.

Radiation resistance of materials regards mainly the changes of the scintillator performance as a consequence, for instance, of colour centre creation. This effect is particularly important in high-energy physics experiments, but it has to be taken into account also in some medical imaging techniques.

9.1.1.7   Density and Effective Atomic Number

To maximise the detection of X- and gamma rays using a scintillator, the crystal should be characterised by a relative high-density ρ and high effective atomic number Z_{eff}. Considering the interaction through the photoelectric effect, the stopping power appears to be proportional to $\rho \cdot Z_{eff}^{3-4}$; therefore, the higher the density and the effective atomic number are, the higher the absorption probability will be.

9.1.2   Scintillation Materials and Application in Nuclear Medicine

After the discovery of X-rays in 1895 by Wilhelm Conrad Roentgen [5], the search for materials to convert this new radiation into visible light started immediately for coupling them with sensitive photographic film-based detectors. CaWO$_4$ phosphor powder was soon introduced in practice, and it was dominantly used until the 1970s, when new and more efficient materials based on Tb-doped oxysulphides (Y$_2$O$_2$S, La$_2$O$_2$S, Gd$_2$O$_2$S) appeared, and their potential for X-ray detection was recognised. Today, the scintillator employed most frequently in X-ray intensifying screens (static imaging) is Gd$_2$O$_2$S:Tb, characterised by an emission spectrum centred at 545 nm due to the Tb^{3+} 4f–4f transition.

In the late 1940s, due to the need to detect and monitor higher-energy X-rays and gamma rays, the interest in the development of single-crystal inorganic scintillators increased. An important family of inorganic scintillators is the alkali halides; among these crystals, NaI:Tl, introduced by Hofstadter in 1948 [6], was the most popular scintillator used in the detectors in the following years because of its high light yield ($\sim 40{,}000$ photons/MeV). The emission spectrum of NaI:Tl, due to the $6p$–$6s$ transition of Tl^+, is centred at approximately 410 nm, and it matches the sensitivity curve of common PMTs. Single crystals of NaI:Tl are grown in ovens under carefully controlled temperature conditions, starting from molten sodium iodide to which a small amount of thallium (0.1–0.4 mol%) has been added [7]. Large crystals can be grown relatively easily and at low cost.

Since NaI:Tl is relatively dense ($\rho = 3.67 \, g/cm^3$) and has a relatively high effective atomic number ($Z_{eff} = 51$), it is a good absorber and efficient detector of penetrating radiations, such as X- and gamma rays in the 50- to 250-keV energy range, where the predominant mode of interaction is by photoelectric absorption. At higher energy (>250 keV), the dominant mechanism of interaction is by Compton effect, and larger crystal volumes are required for adequate detection efficiency [7]. As mentioned, sodium iodide is hygroscopic; consequently, hermetic sealing is required.

NaI:Tl has long been the standard scintillator of gamma camera systems. Originally, these systems utilised monolithic crystal plates in the form of disks with a diameter up to 50 cm and a thickness in the range of 6–12 mm. Modern systems are square or rectangular, up to about $50 \times 60 \, cm^2$ with a crystal plate up to 25 mm thick. Recently, the new scintillators $LaCl_3$:Ce and $LaBr_3$:Ce have been introduced [8, 9], and their development can be of interest for the application in gamma cameras. They are indeed characterised by high light yield, fast response, and good time and energy resolution.

Two other scintillators of the alkali halide family are CsI:Na and CsI:Tl [6, 10]. The first is a good alternative for NaI:Tl in many standard applications because it has high light output ($\sim 85\%$ of that of NaI:Tl) and an emission in the blue spectral region (peak at 420 nm), coinciding with the maximum sensitivity of most popular PMTs with bialkali photocathodes. The second, CsI:Tl, is characterised by an emission spectrum shifted at higher wavelength that peaks at about 550 nm.

Columnar screens of CsI:Na have long been mounted inside the image intensifier-based system used in fluoroscopy to obtain real-time moving images of the internal structures of a patient. Actually, these systems are bulky, and more practical devices have been devloped to efficiently replace these detectors [11]. In particular, the introduction of flat-panel detectors allows the setup of a smaller and more efficient image intensifier. Column-shaped CsI-based scintillators of smooth structure (diameter $\sim 3 \, \mu m$, length > 0.5 mm) can be obtained by vapour deposition and directly deposited over a large-area matrix of amorphous silicon photodiodes. In this case, CsI:Tl is preferred instead of CsI:Na since it is hardly hygroscopic in comparison with CsI:Na, and its emission spectrum matches the quantum efficiency curve of amorphous silicon well.

Two main defects characterise CsI:Tl crystals; the first is a significant afterglow (more than 2% of the maximum scintillation intensity is emitted after ~ 3 ms), and the second is the radiation damage, resulting in a progressive increase of the light output with absorbed dose. Both these defects can be explained and modelled considering the presence of deep traps in the energy-band model of the crystal, where the charge carriers are trapped [12]. Although CsI:Tl crystals have been used in the past in X-ray computed tomographic (CT) systems, the presence of these defects limits the use of this scintillator in modern CT systems characterised by a higher rotating frequency and, consequently, by the need for fast scintillators that do not show afterglow.

Besides NaI:Tl and CsI:Tl crystals, $Bi_4Ge_3O_{12}$ (BGO), introduced by Weber and Monchamp in 1973 [13], also became a widespread scintillator and is often used as a standard scintillator for comparison with newly developed material [1]. BGO is an intrinsic material (self-activated), and it does not need the addition of an external activator like the alkali halides. The emission spectrum, related to the 6s–6p transition of Bi^{3+}, is centred at approximately 480 nm. The main characteristics of BGO are the high density ($\rho = 7.1$ g/cm^3) and the high effective atomic number ($Z_{eff} = 74$), which make this crystal an efficient scintillator for positron emission tomographic (PET) imaging.

An important limit of BGO is the long scintillation decay time ($\sim$300 ns), much longer than the decay time of more recently developed Ce-doped orthosilicate-based scintillators, such as Gd_2SiO_5:Ce (GSO:Ce) [14] and Lu_2SiO_5:Ce (LSO:Ce) [15]. Both these scintillators are characterised by a relatively high density ($\rho = 7.4$ g/cm^3 for LSO and $\rho = 6.7$ g/cm^3 for GSO) and high effective atomic number ($Z_{eff} = 66$ for LSO and $Z_{eff} = 59$ for GSO); consequently, they have good efficiency detection. The emission spectrum is centred at about 420 nm for both crystals, and the scintillation decay time is equal to approximately 40 ns for LSO:Ce and to 60 ns for GSO:Ce. Because of these properties, both these crystals are of interest in PET imaging systems, and large single crystals are commercially available from industrial producers, even if their production is expensive due to their high melting point ($\sim$2,000°C), costly raw material, and demanding technology [1].

Another interesting candidate to replace BGO in PET systems is $LuAlO_3$:Ce (LuAP:Ce) [16]; this scintillator is characterised by a higher light yield and a significantly faster response than BGO (decay time of $\sim$18 ns). It should be noted that, despite the interesting properties of lutetium in the scintillation matrix, this element has a natural radioisotope ([176]Lu, half-life 3.78×10^{10} years, natural abundance 2.59%), which contributes to increase the background signal (count rate of $\sim$300 s^{-1} cm^{-3}).

Recently, a further class of material was introduced: scintillation optical ceramics. They have been under development as an alternative to the single crystals to provide bulky optical elements when single crystals cannot be prepared or their production is too expensive [1]. Advanced ceramic technologies have greatly developed mainly within the last one to two decades; the introduction of co-doping to reduce the afterglow led to mature materials with suitable characteristics for use in CT medical imaging. Examples of ceramic scintillators used in X-ray CT are Gd_2O_2S:Pr, Ce, F [17]; Gd_2O_2S:Pr [18]; and $Y_{1.34}Gd_{0.60}O_3$:(Eu, Pr)$_{0.06}$ [19]. These polycrystalline ceramic scintillators are characterised by interesting scintillation properties, homogeneity, and good mechanical stability. Moreover, their emission spectra match well with the sensitivity of silicon diodes.

In ceramics based on Gd_2O_2S (GOS), scintillation arises from the 4f–4f transition in the Pr^{3+} dopant ions. The decay time is on the order of 3–4 μs, and the afterglow can be strongly reduced by properly co-doping with cerium and fluorine. In $Y_{1.34}Gd_{0.60}O_3$:(Eu, Pr)$_{0.06}$, also known as YGO, scintillation arises from the 4f–4f transition in Eu^{3+}, and the afterglow is reduced by co-dopant ions Pr^{3+} and Tb^{3+}.

New optical ceramics like Lu_2O_3:Eu, Tb, and hafnates ($BaHfO_3$) are, in principle, of interest also in PET applications, provided that better results can be achieved for light yield, fast response time, and energy resolution [11].

The main physical characteristics and scintillation properties of the various compounds mentioned are summarised in Table 9.1 [1–3, 20, 21].

9.2 Semiconductor Detectors

9.2.1 Basic Properties of Semiconductors

Semiconductors are a group of materials between insulators and metallic conductors; they are characterised by an energy gap between the valence and conduction bands typically lower than 3 eV. At low temperatures, all valence electrons remain bound in the lattice; at higher temperatures, thermal vibration may break the covalent bond, and a valence electron may become a free electron, leaving behind a free place or hole. Both the electron and the hole (to be filled by a neighbouring electron) are available for conduction.

The probability per unit time that an electron–hole pair is thermally generated is given by

$$p(T) = CT^{3/2}e^{(-E_g/2kT)}$$

where T is the absolute temperature, E_g is the band-gap energy, k is the Boltzmann constant, and C is a proportionality constant specific for the material [4]. In an intrinsic semiconductor, the equilibrium that is established by the thermal excitation of electrons from the valence to the conduction band and their subsequent recombination leads to equal numbers of electrons in the conduction band and holes in the valence band: $n_i = p_i$ (intrinsic carrier densities). Semiconductor materials are mostly divided into two large classes: elemental semiconductors (group IV of the periodic table) such as silicon, germanium, diamond, and so on and compound semiconductors like IV–IV (SiC), III–V (GaAs, InP, InSb, GaN) and II–VI (CdTe, ZnSe, ZnS, etc.).

Impurities can be introduced into the volume of the semiconductor material, forming an extrinsic semiconductor. Doping can modify considerably the electrical conduction properties of an intrinsic semiconductor. There are two types of impurities: donors or n-type materials and acceptors or p-type materials.

Regarding donors or n-type materials, the atoms of these materials have one surplus electron, which is rather loosely bound in the crystal structure and therefore contributes readily to the negative electron conduction. Therefore, in an n-type semiconductor electrons are the majority carriers. Typical donor materials are antimony (Sb), arsenic (As), and phosphorus (P).

Table 9.1 Characteristics of inorganic scintillators for medical imaging [1–3, 20, 21]

Scintillator	Density (g/cm³)	$\rho \cdot Z_{eff}^4$ (10⁶)	Attenuation length at 511 keV (mm)/Prob. Phot. Eff. (%)	Hygroscopicity	Light yield (photons/MeV)	Decay time (ns)	Afterglow (% after 3 ms/100 ms)	Emission maximum (nm)
NaI(Tl)	3.67	24.5	29.1/17	Yes	41,000	250		410
CsI(Na)	4.51	38	22.9/21	Yes	40,000	630		420
CsI(Tl)	4.51	38	22.9/21	Slightly	66,000	800	>2/0.3	550
Bi₄Ge₃O₁₂(BGO)	7.1	227	10.4/40	No	9,000	300		480
Lu₂SiO₅:Ce (LSO)	7.4	143	11.4/32	No	26,000	40		420
Gd₂SiO₅:Ce (GSO)	6.7	84	14.1/25	No	8,000	60		440
Gd₂O₂S:Tb	7.3	103	12.7/27	No	60,000	1×10^6		545
Gd₂O₂S:Pr,Ce,F	7.3	103	12.7/27	No	35,000	4×10^3	<0.1/<0.01	510
Gd₂O₂S:Pr	7.3	103	12.7/27	No	50,000	3×10^3	0.02/0.002	510
LuAlO₃:Ce (LuAP)	8.3	148	10.5/30	No	12,000	18		365
CaWO₄	6.1	89	13.6/32	No	20,000	6×10^3		420
Y₁.₃₄Gd₀.₆₀O₃:(Eu, Pr)₀.₀₆	5.9	44	17.8/16	No	42,000	1×10^6	4.9/<0.01	610
Lu₂O₃:Eu,Tb	9.4	211	8.7/35	No	30,000	$>10^6$	>1/0.3	611
BaHfO₃:Ce	8.4	142	10.6/30	No	10,000	25		400
LaCl₃:Ce	3.86	23.2	27.8/14	Yes	46,000	25 (65%)		330
LaBr₃:Ce	5.3	25.6	21.3/13	Yes	61,000	35 (90%)		358

For acceptors or p-type materials, the atoms lack one electron in the crystal structure, and they tend to attract an electron, contributing readily to the positive or hole conduction. Therefore, in a p-type semiconductor holes are the majority carriers. Typical acceptor materials are aluminium (Al), boron (B), gallium (Ga), and indium (In).

In extrinsic materials, certain energy levels exist within the forbidden gap between the valence and conduction bands; consequently, the energy required to raise charge carriers from these levels to the conduction state is lower than the energy required to obtain conduction in an intrinsic semiconductor [22].

Semiconductors have properties that make them suitable for the detection of ionising radiation [23]. Semiconductor detectors based on silicon (Si) and germanium (Ge) are used extensively as charged particle and X- and gamma ray spectrometers in physics [4]; their principal application in nuclear medicine is for assessment of radionuclide purity. Miniature gamma ray probes, based on cadmium telluride (CdTe) or cadium zinc telluride (CZT), have been also implemented for application in small nuclear medicine counting and imaging devices and for surgical use [7].

Moreover, due to the small band-gap, semiconductor devices represent a valid alternative to the photomultipliers for the detection of the optical radiation emitted by scintillators. In this case, the small size and the mechanical properties of semiconductor materials may represent an important advantage exploited in the design of several imaging systems. The following sections present a short overview of the basic optical radiation semiconductor detectors.

9.2.2 Photodetection with Semiconductors

Photodetection in semiconductors is achieved following the creation of electron–hole pairs under the action of light. When a semiconductor is illuminated by photons of an energy greater than or equal to its band-gap, photoelectric effects lead to the release of electrons from the valence band into the conduction band, where the free charges may travel long distances across the crystal structure under the influence of an intrinsic or externally applied electric field. Similarly, the holes left in the valence band contribute to electrical conduction by moving from one atom site to another under the effect of the electric field. Therefore, the separation of electron–hole pairs generated by the absorption of light gives rise to a *photocurrent*, defined as the fraction of the photogenerated free charge carriers collected at the edges of the material by the electrodes of the photodetecting structure [24].

9.2.2.1 Bulk Semiconductor Photodetectors

A bulk semiconductor photodetector consists of a homogeneous material working as a photoconductor. In photoconductors, a certain number of conduction electrons is available at room temperature so that, in response to a bias voltage, a small current flows even without irradiation. The interaction of photons with the material leads to

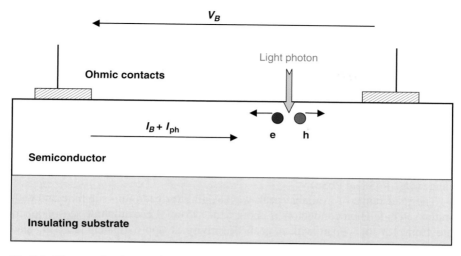

Fig. 9.2 Diagram of a photoconductor detector

the absorption of energy, which is used to release additional electrons either from the valence state in intrinsic photoconductors or from dopant energy states in extrinsic photoconductors. The incident radiation therefore generates additional mobile charge carriers, effectively lowering the resistance or increasing the conductance of the material.

A diagram of a photoconducting device is shown in Fig. 9.2: A fixed voltage V_B is applied between the two ohmic contacts; consequently, a bias current I_B flows through the semiconductor layer following Ohm's law. When the active optical surface is illuminated, the photogenerated charges originate a photocurrent I_{ph}, which is added to the bias current, effectively increasing the conductivity of the device.

The spectral responsivity of a photoconductor is given by

$$s(\lambda) = \frac{I_{ph}}{\Phi} = \frac{(1 - R) \cdot G \cdot \eta \cdot q \cdot \lambda}{hc}$$

where I_{ph} is the photocurrent, Φ is the incident flux, R is the reflectance of the front surface, η is the quantum efficiency, q is the electron charge, λ is the wavelength of the incident photon, h is the Planck constant, c is the speed of light, and G is the photoconductive gain. The photoconductive gain is given by

$$G = \frac{\tau}{T_r}$$

where τ is the lifetime of the charge carriers, and T_r is the travel time between electrodes. The gain G is directly proportional to the bias voltage.

If the quantum efficiency η is independent of the wavelength, the spectral responsivity function increases linearly with wavelength to a maximum value, then it decreases rapidly. The maximum occurs at

$$\lambda_{\max} = \frac{hc}{E_V}$$

where E_V is the energy gap between the state of the valence electron and the conduction electron. Photocurrent is indeed produced only if the photon energy $E_{ph} = h\nu = \frac{hc}{\lambda}$ is larger than the energy gap E_V.

Depending on their spectral responsivity function, photoconductive detectors can be divided into photoconductive detectors for the visible wavelength range (e.g., CdS and CdSe) and photoconductive detectors for the near-infrared wavelength range (e.g., PbS and PbSe).

The total range of available peak wavelengths for cadmium sulphide and cadmium selenide photoconductors is about 515–735 nm. Lead sulphide sensors operate from 1.3 to 3.2 μm with a peak sensitivity at approximately 2.2 μm; lead selenide sensors operate from 1.3 to 5.2 μm with a peak sensitivity near 4.2 μm [22].

9.2.2.2 Junction Semiconductor Photodetectors

The most important electronic structure obtained by joining extrinsic semiconductors of opposite doping is the p–n junction. In this structure, a voltage applied in one direction, the so-called forward direction, causes a significantly larger current than that obtained by applying the same voltage in the other direction, the so-called reverse direction. In junction photodiodes, the two materials forming the junction are selected so that, even without external bias, a potential develops across the junction. This potential depletes the vicinity of the junction of charge carriers, resulting in high resistance. Radiation incident on the junction generates electron–hole pairs that are separated by the internal potential and causes a current through the external circuit or, if an external reverse bias is applied, increases the conductance so that a current flows in response to this bias. Several junction devices therefore may be used either as photovoltaic or as photoconductive detectors; they may be used unbiased, generating a voltage or a current that is measured by a suitable electronic device, or they may be used similar to photoconductors by applying an external reversed bias voltage and measuring the change of resistance caused by the incident radiation.

Different junction photodetector structures exist. The simplest configuration is the p–n photodiode, which is formed by diffusing a thin layer of acceptor or p-type impurity into the surface of a donor or n-type doped slice of silicon so that in this thin layer the concentration of acceptors exceeds that of donors. Similarly, n–p photodiodes are made by diffusing an n-type impurity into a p-type doped material. In both cases, the junction is the region where the n-doped and p-doped materials meet. A schematic diagram of a planar diffused p–n silicon junction photodiode is shown in Fig. 9.3.

An additional junction photodetector structure is the Schottky barrier photodiode. It is obtained by deposition of a metal film on a semiconductor surface in such a

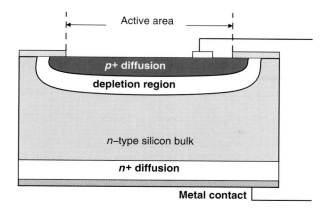

Fig. 9.3 Schematic diagram of a planar diffused p–n silicon junction photodiode

Table 9.2 Materials commonly used to produce junction semiconductor photodetectors

Material	Wavelength range (nm)
Silicon (Si)	190–1,100
Germanium (Ge)	400–1,700
Indium gallium arsenide (InGaAs)	800–2,600

way that no interface layer is present. The barrier thickness depends on the impurity dopant concentration in the semiconductor layer. The electron–hole pairs generated by the incident radiation are separated by the potential barrier between the metal and the semiconductor.

In a p–n photodiode, electron–hole pairs created in the depletion region contribute to the electric signal, but those released outside this region travel by diffusion or simply recombine. To expand the depletion region to include carriers released by a broader range of photon wavelengths, the p–i–n structure can be used. A p–i–n photodiode is a planar diffused diode consisting of a single crystal having an intrinsic (undoped or compensated) region sandwiched between p- and n-type regions. A bias potential applied across the detector depletes the intrinsic region of charge carriers, constituting the radiation-sensitive detector volume.

When the reverse bias voltage established at the terminal of a p–i–n structure is sufficiently increased, photogenerated carriers are accelerated enough to generate further carriers by ionisation; in this condition, multiplication of charge carriers gives rise to an internal gain mechanism. This effect is exploited in the avalanche photodiodes, in which the levels of doping are generally adjusted to high values above 10^{18} cm^{-3} to maximise the intrinsic electric field of the junction [24].

The materials commonly used to produce junction semiconductor photodetectors, together with the wavelength range that characterises their spectral responsivity curve, are reported in Table 9.2.

References

1. Nikl M. Scintillation detectors for x-rays. Meas. Sci. Technol. 17, R37–R54 (2006).
2. van Eijk C.W.E. Inorganic-scintillator development. Nucl. Instr. Meth. Phys. Res. A, 460, 1–14 (2001).
3. Weber M.J. Scintillation: mechanism and new crystals. Nucl. Instr. Meth. Phys. Res. A, 527, 9–14 (2004).
4. Knoll G.F. Radiation detection and measurements. (Wiley, New York) (2000) ISBN 0-471-07338-5.
5. Roentgen W.C. On a new kind of rays. Science 3, 227–231 (1986).
6. Hofstadter R. Alkali Halide Scintillation Counters. Phys. Rev. 74, 100–101 (1948).
7. Cherry S.R., Soreson J.A., Phelps M.E. Physics in Nuclear Medicine (Saunders, Philadelphia) (2003) ISBN-13: 978–0–7216–8341–6; ISBN-10: 0–7216–8341-X.
8. van Loef E.V.D., Dorenbos P., van Eijk C.W.E., Krämer K., Güdel H.U. High-energy resolution scintillator: Ce^{3+} activated $LaCl_3$. Appl. Phys. Lett. 77, 1467–1468 (2000).
9. van Loef E.V.D., Dorenbos P., van Eijk C.W.E., Krämer K., Güdel H.U. High-energy resolution scintillator: Ce^{3+} activated $LaBr_3$. Appl. Phys. Lett. 79, 1573–1575 (2001).
10. Tyrrel G.C. Phosphors and sintillators in radiation imaging detectors. Nucl. Instr. Meth. Phys. Res. A, 546, 180–187 (2005).
11. Van Eijk C.W.E. Inorganic scintillators in medical imaging. Phys. Med. Biol. 47, R85–R106 (2002).
12. Wiekzorec H. and Overdick M. Afterglow and hysteresis in CsI:Tl. Proc. 5th International Conference on Inorganic Scintillators and Their Application (Moscow: M.L. Lomonosov Moscow State University), 385–390 (2000).
13. Weber M.J. and Monchamp R.R. Luminescence of $Bi_4Ge_3O_{12}$: spectral and decay properties. J. Appl. Phys. 44, 5495–5499 (1973).
14. Takagi K. and Fukazawa T. Cerium-activated Gd_2SiO_5 single crystal scintillator. Appl. Phys. Lett. 42, 43–45 (1983).
15. Melcher C.L. and Schweitzer J.S., 1992 Cerium-doped lutetium oxyorthosilicate: a fast, efficient new scintillator. IEEE Trans. Nucl. Sci. 39, 502–505 (1992).
16. Minkov B.I. Promising new luthetium based single crystals for fast scintillation. Funct. Mater. 1, 103–105 (1994).
17. Yamada H., Suzuki A., Uccida Y., Yoschida M., Yammoto H. A scintillator Gd_2O_2S:Pr,Ce,F for x-ay computed tomography. J. Electrochem. Soc. 136, 2713–2720 (1989).
18. Rossner W., Ostertag M., Jermann F. Properties and applications of gadolinium oxysulfide based ceramic scintillators. J. Electrochem. Soc. 98–24, 187–194 (1995).
19. Greskovich C. and Duclos S. Ceramic scintillators. Ann. Rev. Mater. Sci. 27, 69–88 (1987).
20. Liu B. and Shi C. Development of medical scintillator. Chin. Sci. Bull. 47, 1057–1063 (2002).
21. Novotny R. Inorganic scintillators – a basic material for instrumentation in physics. Nucl. Instr. Meth. Phys. Res. A, 537, 1–5 (2005).
22. Budde W. Physical Detectors of Optical Radiation. Optical Radiation Measurements Vol. 4 (London Academic Press, New York) (1983) ISBN 0–12–304904–0.
23. Lutz G. Semiconductor Radiation Detectors. Device Physics (Springer, Berlin) (1999) ISBN 3–540–64859–3.
24. Omnes F. Optoelectronic Sensors. Chapter 1 (Didier Decoster and Joseph Harari, Polytech'Lille, France) (2009) ISBN: 9781848210783.

Chapter 10
New Trends in Detectors for Medical Imaging

Gabriela Llosá and Carlos Lacasta

10.1 Introduction

The improvement of medical imaging instrumentation is based both on the enhancement of the performance of the detectors employed in their construction and on the development of novel techniques and methods that allow full exploitation of the advantages of the new instrumentation.

Most detectors in medical imaging are composed of scintillator crystals coupled to photodetectors. Two main ways are investigated for advanced performance. One is the improvement of the response of each of the components commonly employed, for instance, crystals with higher light yield, higher stopping power, and shorter decay time and position-sensitive photodetectors with high photon detection efficiency (PDE) and small dead areas. Silicon photodetectors, avalanche photodiodes (APDs), and silicon photomultipliers (SiPMs) have improved their performance considerably since 2005, and their use in different applications is rapidly increasing.

Another approach that is gaining interest is that of alternative types of detectors, for example, solid-state detectors for particular applications in positron emission tomography (PET), single-photon emission computed tomography (SPECT), or computed tomography (CT). Currently, these alternatives do not offer a substantial improvement that makes it possible to replace traditional detectors based on scintillators and photodetectors, but they can improve performance significantly in some aspects (e.g., in spatial resolution and in determined applications). In both cases, the use of detectors and electronics initially developed for high-energy physics, such as solid-state detectors or gas detectors, fast electronics, dedicated application-specific integrated circuits (ASICs) and data acquisition systems, and the always increasing common interaction between both fields have been beneficial for the field of medical imaging.

G. Llosá (✉) and C. Lacasta
Edificio Institutos de Investigación, IFIC Instituto de Física Corpuscular, CSIC-Universitat de
València, Ap. Correos 22085, 46071 Valencia, Spain
e-mail: gabriela.llosa@ific.uv.es

M.C. Cantone and C. Hoeschen (eds.), *Radiation Physics for Nuclear Medicine*,
DOI 10.1007/978-3-642-11327-7_10, © Springer-Verlag Berlin Heidelberg 2011

The development of detectors with improved performance also makes the implementation of novel techniques possible that significantly contribute to improve the performance and, in turn, impose new challenges for further development of the detectors. Among these is time-of-flight (TOF) PET, which uses the difference in the arrival time of the two photons in two opposite detector heads to restrict the determination of the point of origin of the photon pair to a small region instead of a line of response (LOR), reducing the signal-to-noise ratio (SNR) in the reconstructed images. This technique is possible thanks to the development of fast scintillator crystals and photodetectors.

New detector geometries are investigated for PET and SPECT, with particular attention on the determination of the depth of interaction (DOI), that is, the depth in the detector head at which the interaction takes place, to reduce the parallax error and to improve the spatial resolution in PET applications.

Multimodality is the combination of different imaging techniques, which leads to enhanced diagnostic value of the resulting images. It requires high performance in each of the modalities, and it generally demands the modification of the detectors for a full profit from the combination.

The expansion of new fields such as proton and ion therapy also increases requirements for imaging detectors, dosimetry, and beam monitoring.

It must be taken into account that the developments in detector construction must always run parallel with developments in data processing and analysis and in image reconstruction algorithms, which allow full profitability from the benefits of the novel instrumentation and techniques.

This chapter gives an overview of the main research lines in detector development that have been of interest to the scientific community in recent years in some medical imaging fields. First, a short description of the detectors currently employed and principal areas of development is given. Then, the main trends in different medical imaging techniques in PET, SPECT, and CT applications are briefly summarized.

10.2 Detector Development

In general, in all medical imaging fields, detectors with higher resolution, higher sensitivity and faster timing performance are desired. Currently, most applications employ inorganic scintillator crystals coupled to photomultiplier tubes (PMTs), which continue to be the main area of development. Scintillators and PMTs with improved performance are investigated as are new types of photodetectors. Also, the use of other types of detectors, such as solid-state or gaseous detectors, instead of the traditional detectors can lead to improved performance of medical imaging devices.

10.2.1 Scintillator Crystals

For many years, there has been a continuous search for new scintillator crystals, and a large number of new types of scintillators with different properties

Table 10.1 Main properties of scintillator crystals used in medical imaging.

	Density (g/cm^3)	Light yield (photons/MeV)	Decay time (ns)
NaI	3.67	41000	260
CsI	4.51	40000	630
BGO	7.1	9000	300
LSO	7.4	26000	40
LaBr$_3$	5.3	61000	∼20

have been fabricated and employed in medical applications, such as sodium iodide (NaI), cesium iodide (CsI), bismuth germanate oxide (BGO), lutetium orthosilicate (LSO) or lutetium yttrium orthosilicate (LYSO), gadolinium oxyorthosilicate (GSO) (Gd_2SiO_5:Ce), yttrium aluminum perovskite (YAP), lutetium aluminum perovskite (LuAP), Barium fluoride (BaF_2), Lutetium iodide (LuI_3), lanthanum chloride ($LaCl_3$), and lanthanum bromide ($LaBr_3$), among many others. However, the ideal scintillator has not yet been found, and the most suitable scintillator for each application must be selected based on the requirements of the application. A review of the main scintillation crystals employed in medical imaging can be found in [1].

This section describes briefly the advantages, drawbacks, and uses of those scintillators that have been mainly used in medical imaging (NaI, BGO) and those that have currently the highest interest for the medical imaging community since they are used in most of the novel research lines (LSO/LYSO and $LaBr_3$). The main properties of these scintillators (density, light yield and decay time) are summarized in table 10.1.

The fundamental properties of a scintillator crystal are high light yield, high rise time, short decay time, and high photoabsorption probability for the energies considered in a given application to stop the photons in just one interaction. A high light yield results in better spatial and energy resolution. The fast rise time leads to better timing properties, which contribute to reduce noise and random events, and a short decay time results in reduced dead time and pileup and therefore better count rate performance and greater efficiency.

For many years, NaI, CsI, and BGO have been the main scintillators employed. Due to its high light yield, NaI has been widely used for many applications, including both PET and SPECT. However, given its relatively low stopping power, it is better suited for SPECT applications, in which photons generally have lower energies. BGO has a high stopping power; therefore, since its development it has been employed largely for PET applications. Its main drawback is its low light yield.

Although its production is more expensive than that of the NaI and BGO, LSO (or LYSO) is now the scintillator most widely employed for PET applications given its higher light yield and relatively short decay time. This scintillator is currently employed in many research applications, including TOF PET. However, the presence of the radioactive isotope Lu-176 in the crystal, with a natural abundance of 2.59%, causes a radioactive background that makes it less indicated for SPECT applications.

LaBr$_3$ has become a promising scintillator given its high light yield and short decay time. Its main drawback is that it does not have high stopping power. Several groups are now investigating this new scintillator for SPECT [2] and PET [3], including TOF techniques. With high PDE photodetectors, detectors employing LaBr$_3$ have achieved an energy resolution close to that of solid-state detectors (6.5% at 511 keV). Its lower stopping power compared to LSO must be compensated by increasing the crystal thickness to achieve similar efficiency. Its fast timing properties lead to excellent timing resolution that is being investigated to replace LSO in TOF PET applications, improving the timing performance.

In any case, no scintillator with probability of photoelectric absorption greater than 50% for 511-keV gamma rays has yet been found; therefore, multiple interactions can take place in the crystal before the photons are absorbed. If all the interactions take place in the same crystal, this effect, which increases with crystal thickness, degrades the intrinsic spatial resolution. In segmented detectors, algorithms can be applied to determine the first interaction and improve the spatial resolution.

The crystal geometry and arrangement for a better spatial resolution and efficiency are also an active area of research. Different geometries with pixelated and continuous (monolithic) crystals and crystal configurations for DOI determination in PET applications are discussed in Sect. 10.3.1.

10.2.2 Photodetectors

Photodetectors are mainly employed to detect the optical photons produced by the interaction of the gamma rays in the scintillator crystals. The main requirements of a photodetector are high PDE and a small dead area to detect the maximum possible number of photons. They must have fast timing resolution, negligible in comparison to the scintillator crystal coupled to it. High gain leads to a good SNR, making the use of low-noise electronics unnecessary. Compact photodetectors are preferred, particularly in applications in which small detectors are required, and a small dead area enhances the sensitivity. A low bias voltage results in lower power consumption and less heating.

The quantum efficiency (QE) of a photodetector is the probability that an optical photon that reaches the active surface of the photodetector interacts in it, emitting a photoelectron in PMTs or creating an electron–hole pair in solid-state photomultipliers, which can produce a signal. The QE depends on the photon wavelength. The PDE is the probability that an optical photon produces a signal, taking into account the QE and other effects, such as the collection efficiency in PMTs, or the probability that the carriers trigger an avalanche in silicon photodetectors (APDs and SiPMs).

In addition to the advances in conventional PMTs, silicon photodetectors have experienced considerable improvement in performance, and their use for different applications is rapidly increasing.

10.2.2.1 Photomultiplier Tubes

PMTs have been the principal photodetectors in physics and medical imaging for many years, and they are still employed in most of the experimental and commercial systems. They have high gain ($\sim 10^6$–10^7) and low noise, they are fast, and they have stable operating conditions. However, they also present some disadvantages: Their QE is not high (generally around 20%), they are bulky, and their operating voltage is high (up to a few thousands of volts). The fact that they are sensitive to magnetic fields has been the main problem for the combination of PET and magnetic resonance imaging (MRI) modalities, which are described in Sect. 10.3.1.

A fundamental research line, common to all fields in physics, is the improvement of the QE of the PMTs. PMTs with QE above 40% with bialkali or GaAsP photocathodes have been produced [4]. In this type of detector, the probability that the photoelectrons emitted in the photocathode give a signal is high for photons interacting at the center of the photodetector; thus, the PDE is close to the QE. Another important request is the development of fast devices. Fast PMTs have allowed the development of TOF PET detectors, and the possibility of further improving these devices is under investigation [5].

Since their development, the use of position-sensitive PMTs (PSPMTs), also known as multianode PMTs, has been widely adopted for medical imaging applications. These devices consist of multiple systems (cathode–dynode chain–anode) acting as independent systems. Several manufacturers produce these devices, with an increasing number of channels per unit area, such as Hamamatsu H7546 with 64 channels (Fig. 10.1) or H9500 with 256 (16 × 16) channels in 52 × 52 mm. Low dead space in the edges is also important to allow placement of several photodetectors side by side to form a detector unit of larger area. These photodetectors are employed with both pixelated and continuous crystals, either multiplexing the output signals to reduce the number of channels, which is the most common approach, or reading out all output channels, which is gaining interest thanks to the possibility of employing ASICs.

10.2.2.2 Silicon Photodetectors

Silicon photodetectors have become a sound alternative to vacuum devices thanks to the considerable improvements achieved since 1999. These devices have a high QE for optical photons (up to above 90%); they are compact and rugged and potentially fast. In addition, they are insensitive to magnetic fields, which makes them the ideal photodetectors for applications such as the combination of PET and magnetic resonance (MR), which is explained in Sect. 10.3.1. On the other hand, their main drawbacks are their dark rate due to thermal generation of carriers, which is higher than for PMTs but can generally be avoided by setting a high threshold, and the dependence of their characteristics on temperature.

Silicon diodes can be used as photodetectors. The optical photons interact in the detector, generating carriers that produce an electric signal, but the low amount of

Fig. 10.1 Position-sensitive photomultiplier tube (PSPMT) Hamamatsu H7546 with 64 channels

carriers produced results in a small signal, and since the device has no internal gain, the SNR is low. For this reason, they are not generally used for this purpose.

APDs are silicon detectors operated close to the breakdown point; they have a structure that creates a high electric field region in the device. The carriers generated by the optical photons reach this region, in which they are accelerated. The collision of the carriers with the lattice can ionize the atoms and create new carriers, producing an avalanche. APDs thus have an internal gain, which is generally between 50 and 2,000. The signal produced is proportional to the number of optical photons that reach the detector. The bias voltage is commonly between 500 and 1,500 V; in general, the higher the reverse voltage is, the higher the gain will be.

APDs produced recently have improved their performance significantly with respect to the first samples available, mainly in terms of operation stability and showing an increase of the gain up to 1,000–2,000, and they are currently employed in many experimental and commercial applications, principally in PET [6, 7] and SPECT [8] and in PET/MRI applications [9].

APDs have a high QE, and the probability that the carriers trigger the avalanche is also high in these photodetectors. Their main disadvantage is their low gain, which makes it necessary to use low-noise electronics for the signal readout, which also degrades the timing properties. In addition, the statistical fluctuations of the gain are high, and as a result, the energy resolution is not as good as expected. This effect is described by the excess noise factor F. The noise due to the multiplication process at a gain M is denoted by $F(M)$ and can be expressed as

$$F(M) = \kappa M + \left(2 - \frac{1}{M}\right)(1 - \kappa),$$

where κ is the ratio of the hole impact ionization rate on that of the electrons. For an electron multiplication device, it is given by the hole impact ionization rate divided by the electron impact ionization rate. $F(M)$ is a main factor, among other things, that limits the best possible energy resolution obtainable.

To have two-dimensional (2D) information on the interaction position of the gamma rays in a crystal, 2D arrays of APD detectors a few millimeters in size are fabricated. Another solution is the fabrication of large-area (up to 64 × 64 mm) position-sensitive APDs, which provide a signal output at the four corners of the device. Combining the four signals, it is possible to identify the independent crystals of a crystal array [8].

Multicell Geiger-mode APDs, commonly known as SiPMs, are a new type of photodetector that has experienced fast development [10]. These detectors consist of a 2D array of tiny structures (typically 100–$2,000/mm^2$) known as microcells. Each microcell is an independent APD that is biased slightly above the breakdown voltage and operates in Geiger mode. The avalanche produced is independent of the number of photons that arrive at the microcell. All the microcells are connected in parallel, and the output of the SiPM is given by the sum of the outputs of all the microcells that trigger. The SiPM output will be proportional to the number of photons that reach the detector if this number is low compared to the number of microcells. For a high number of photons, the SiPM response deviates from linearity.

These detectors have high gain (around 10^5–10^6); therefore, low-noise electronics are not required. They have fast timing properties (about 120 ps full width at half maximum [FWHM] at the single-photoelectron level) and a low operation voltage, typically below 100 V. The PDE for blue light has improved significantly, and current devices have a PDE above 30% for all wavelengths. The PDE in these devices is given by the product of three factors [11]: the QE, the triggering probability (Pt), and the fill factor (FF). The QE is high for optical photons, as in the case of APDs. The Pt is given by a combination of the Pt of electrons and holes, and it will be high if the electrons are the carriers that trigger the avalanche. This depends on the SiPM structure and on the interaction position of the optical photons in the detector, which is given by the wavelength. The fill factor is the ratio of the active to the total area of the microcells.

As in the case of APDs, 2D arrays have been developed (Fig. 10.2). Some manufacturers are developing 2D arrays with large numbers (up to 64) of SiPM elements in a common substrate, with readout lines that take the signals to the edges of the detector, to minimize the dead area between the SiPM elements.

10.2.3   Solid-State Detectors

Solid-state detectors (silicon, CZT [cadmium zinc telluride], and CdTe) are used as intrinsic detectors that directly detect the incoming gamma radiation and produce electrical signals. The main advantages of solid-state detectors are their excellent energy resolution and their high spatial resolution. Thin detectors are fast, but the

Fig. 10.2 Silicon photomultiplier (SiPM) two-dimensional (2D) array from Hamamatsu, composed of 16 SiPM elements 3 × 3 mm size

timing resolution is degraded for thick detectors (1 mm or more). Their stopping power is lower than that of scintillators; therefore, their sensitivity is lower for the same thickness. However, their high spatial and energy resolution can be employed for better determination of scattered events, reducing the amount of events rejected and enhancing the sensitivity. In these detectors, it is also possible to distinguish multiple interactions. An important drawback is the cost, which is higher than that of common scintillators.

These detectors are generally thin in comparison with scintillators, and they are operated in stacks, providing discrete DOI information that is given by the detector in which the interaction has taken place.

10.2.3.1 Silicon Detectors

Silicon detectors have excellent energy resolution for gamma rays (1.5 keV FWHM at 59.5 keV, 2.5% FWHM) [12] and good spatial resolution, but their photoabsorption probability is much lower than that of scintillator crystals (Si 0.2%; LSO 34%; BGO 44% for 511-keV photons). In spite of this, several groups are investigating different possibilities that allow benefitting from their high spatial and energy resolution. Silicon detectors are of particular interest in some PET applications.

The high Compton interaction probability can be useful in small-animal PET. Events in which photons undergo Compton scattering followed by an escape are usually rejected, given that they cannot be distinguished from the events in which the photon scatters in the patient before absorption. However, in small-animal PET the scattering probability in the mouse or rat is low, and the effect of including the scattered events does not degrade the resulting images significantly. Compton events are thus employed to determine the LOR and reconstruct the images [13].

Another possibility that is under study is the use of the silicon detectors to add a second detector ring or partial ring to enhance the spatial resolution. This approach is employed for small-animal PET, including a silicon ring inside a conventional scintillator ring and considering the different types of events. The events in which both photons interact in the silicon ring and are afterward absorbed in the scintillator provide excellent spatial resolution. The events in which one of the photons interacts in the silicon ring and the other is directly absorbed in the scintillator have medium resolution. If both photons interact in the scintillator ring, the detector is a normal PET scanner [12].

The minimizing activities and doses by enhancing image quality in radiopharmaceutical administration (MADEIRA), http://www.helmholtz-muenchen.de/en/madeira/ project also employs a silicon probe consisting of a partial ring of stacked detectors that can be moved to the region of interest to enhance the spatial resolution. A detailed description of this project is given in Chap. 11.

There is also increased interest in silicon detectors in SPECT applications in Compton imaging. This technique is explained in the section on SPECT. The silicon detectors also have applications as detectors for dosimetry and beam monitoring in hadron therapy.

10.2.3.2   CZT and CdTe Detectors

CZT and CdTe detectors have a higher atomic number (Z) than silicon; therefore, they have higher photoabsorption probability, and they can be used in PET applications. However, the stopping power is in general still lower than that of scintillators, and a larger detector thickness is necessary to achieve a similar efficiency. Their high cost and relatively poor timing resolution ($\sim$10 ns) compared to scintillators are the main drawbacks of this type of detector for medical applications. Nevertheless, their high spatial (<1 mm) and energy resolution ($\sim$4% FWHM at 511 keV) makes them attractive detectors for small-animal PET [14, 15].

CZT detectors can also offer some advantages in SPECT applications [16]. Slat collimators have higher sensitivity than parallel hole collimators for the same spatial resolution. However, the amount of scattered photons detected is also higher. The use of CZT detectors with good spatial resolution (about 3% at 140 keV) allows the rejection of scattered photons, making use of these collimators advantageous and resulting in a gain in sensitivity.

The use of these detectors for CT as direct detectors is also under investigation.

10.2.4   Liquid and Gas Detectors

Liquid and gas detectors are not often employed in medical imaging due to their low stopping power for gamma rays. Their main advantages are excellent spatial resolution and the possibility of covering large detector areas at low cost, which partially compensates the loss of sensitivity in comparison with scintillator crystals. However, the large volume also implies higher scattering between modules and random contributions. The field of view (FOV) can also be large in these detectors, avoiding image distortions due to activity outside the FOV.

A PET system with detector modules consisting of multiwire proportional chambers has been employed for small-animal PET, achieving excellent spatial resolution around 1 mm FWHM and with 10% uniformity within the FOV thanks to its DOI determination capability [17].

Liquid xenon detectors are also under investigation for PET applications [18]. An energy resolution better than most scintillator detectors (below 10%) can be achieved by combining the measurement of the collection of the electron drift charge and of the scintillation light detected with photodetectors. In addition to the *xy* coordinates, the depth of interaction can be calculated by measuring the time difference between the light flash in the photodetectors and the charge arrival on the anode plate. Subnanosecond timing resolution can be achieved.

Gas detectors are also employed for dosimetry in hadron therapy [19].

10.2.5 Electronics

As the performance of fast detectors with higher efficiency and count rate capability increases, electronics that can read out and process data in a short time are essential in medical imaging. The development of systems with a high number of readout channels requires the use of ASICs, which have significant advantages in terms of cost, size, power, reliability, and performance. Different groups are working on the development of ASICs adapted to the requirement of detectors employed for many medical applications.

The Medipix detector [33] is an example of detector development for different application fields. A family of photon-counting pixel detectors has been developed by an international collaboration with applications in high-energy physics, astronomy, and medical imaging. MEDIPIX is a CMOS (complementary metal oxide semiconductor) pixel detector readout chip designed to be connected to a segmented semiconductor detector. The semiconductor sensor layer (silicon, GaAs, or CdTe) that detects the incident radiation is bump bonded to the CMOS electronics layer, which counts the number of events on each pixel within a selected energy range, enabling spectroscopic X-ray imaging. In medical imaging, it is mainly used in X-ray and CT applications [20].

The last generation, MEDIPIX3, has 256×256 pixels of $55 \times 55\,\mu m$, with a total active area of 14.08×14.08 mm. This version has the capability of real-time correction of charge sharing, which results in better energy resolution. In addition, multiple counters per pixel allow for continuous readout and up to eight energy thresholds [21].

10.3 Applications

10.3.1 Positron Emission Tomography

The improvement of the performance of PET detectors is addressed to fulfill different requirements depending on the applications. In preclinical (small-animal) detectors, the main goal is to improve the spatial resolution, aiming at a resolution better than 1 mm^3 in the whole FOV. DOI determination is therefore essential in these systems. Timing resolution is not a stringent requirement since the random rates are lower than in larger scanners due to a lower single-event count rate, and TOF PET is not implemented in such small devices. In whole-body images, spatial resolution is not the limiting parameter now. The intrinsic resolution of the detectors can be improved by reducing the crystal size in the pixellated crystals generally used, but other factors, such as photon acolinearity, are significant. A higher sensitivity and especially a better SNR, which can be achieved with TOF techniques, are the main requirements. In dedicated scanners (e.g., for brain or breast imaging), better spatial resolution is also desired. Although TOF PET does not lead to a considerable improvement with current detectors, good timing resolution would significantly improve the random rejection. In all applications, a higher sensitivity is desired.

Most PET detectors are composed of pixellated crystal arrays. In this case, the improvement of the spatial resolution is obtained by reducing the crystal size. However, the cost of manufacturing arrays of small crystals is high, and increasing the number of crystals also increases the dead area, reducing the efficiency. With the development of finely segmented photodetectors (PSPMTs, APDs, SiPMs), the use of continuous crystals has gained renewed interest [22, 23]. In these crystals, the determination of the interaction position is more complicated than in the case of pixelated crystals, and they generally have problems near the edges of the detector. Position reconstruction algorithms with maximum likelihood methods and neural networks are under study and are producing promising results [24].

The use of thick crystals contributes to enhance the efficiency of PET detectors. However, the lack of information about the depth at which the interaction takes place leads to a wrong determination of the LOR. This effect, known as the *parallax error*, gets worse toward the outer part of the FOV (see Fig. 10.3), leading to degradation of the spatial resolution. Research in this area is active, both for pixelated and for continuous crystals. The many different methods investigated to determine the DOI are well described in [25]. There are three main approaches.

One method employed with pixelated crystals is the use of different types of scintillator crystals with different decay times, one on top of the other and coupled to the same photodetector (Fig. 10.4a), or two crystals of the same type with a phosphor layer in between that absorbs the light and reemits it with a different decay time. Pulse shape discriminator techniques allow distinguishing the signals coming from one or another crystal. This method is known as a *phosphor sandwich* or *phoswhich* detector.

Fig. 10.3 Effect of the depth
of interaction (DOI):
uncertainty and parallax error.
There is a difference in the
line of response (LOR) for
photons that interact close to
the detector surface, or deeper
in the detector, while in
practice the same LOR is
assigned if the detector has no
DOI determination capability.
This difference is larger for
detectors at the edges of the
field of view (FOV)

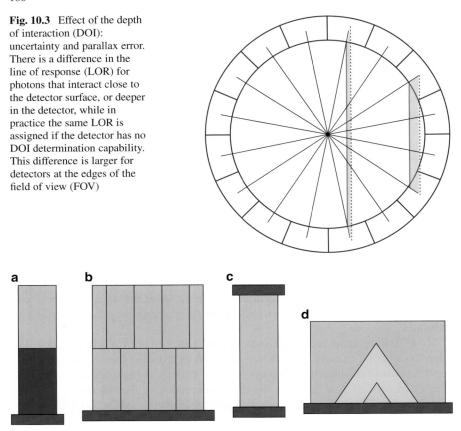

Fig. 10.4 Different methods to determine the depth of interaction (DOI) position. (a) Phoswich detector. (b) Displacement of one crystal layer with respect to another. (c) Photodetector readout on both crystal ends. (d) DOI information obtained from the light pattern in the photodetector

Another method also applied to pixelated crystals is to offset different detector layers or to place absorbing and reflecting materials asymmetrically between the crystals in such a way that the light coming from each of the crystals is detected in a different position in the photodetector (Fig. 10.4b).

A third approach employed both with pixelated and continuous crystals is to place photodetectors at both ends of the crystal. The ratio of the light collected in the two photosensors gives an estimate of the DOI (Fig. 10.4c). In this configuration, at least one of the photodetectors must be thin to allow gamma rays to reach the crystal. Silicon photodetectors are generally employed. Resolutions of 2 mm have been obtained with this method. In the case of continuous crystals, it is also possible to use just one photodetector and to obtain the DOI information from the light pattern in the photodetector (Fig. 10.4d).

With solid-state detectors and with solid-state photomultipliers coupled to scintillator crystals, it is possible to stack several detector layers that act as independent detectors, up to the desired detector thickness. The DOI information is given by the position of the detector in the stack that gives the signal. This method can be employed with both pixellated [7] and continuous crystals [26]. The main drawback is that the number of readout channels increases compared to thick detectors for the same efficiency.

With the development of fast scintillators and photodetectors, the TOF PET technique has become possible. In human (whole-body) PET scanners, the difference in the arrival time of the two coincidence photons Δt can be determined [5] with a given uncertainty. The location of the annihilation event with respect to the midpoint between the two detectors is given by

$$\Delta d = \Delta t \times \frac{c}{2}$$

where c is the speed of light. Thus, a timing resolution of 60 ps would result in a depth resolution of about 1 cm. The noise in the images reconstructed with an appropriate algorithm is substantially reduced, and a better SNR is achieved.

This technique requires scintillators with fast rise time and fast photodetectors with a high PDE. The first TOF PET detectors have been commercialized with LSO crystals and fast PMTs. This combination is still under investigation; studies with LSO and $LaBr_3$ crystals coupled to fast PMTs or SiPMs are taking place. Resolutions below 200-ps FWHM have already been achieved, and even better resolutions are possible with SiPMs and $LaBr_3$ crystals.

The combination of structural and functional imaging modalities leads to a significant improvement in the diagnostic value, as it is clear from the combination of PET/CT, that is replacing PET-only systems. MRI is superior to CT in terms of soft tissue contrast, and no additional radioactive dose is administered to the patient. In addition, the PET and CT images are acquired sequentially, increasing the possibility of errors in the coregistration of the images due to movement of the patient, while PET/MR systems are designed for simultaneous acquisition in both modalities.

The main difficulty for the combination of PET and MR has been the sensitivity to magnetic fields of the PMTs used in PET detectors. Some solutions have been found employing light guides to convey the light from the scintillator crystals to the outside of the magnet, where the PMTs can be operated [27]. Unfortunately, this generally leads to a degradation of the PET performance. Another solution has been found with the use of APDs, which are insensitive to magnetic fields, as photodetectors [9] and has allowed the combination of the two modalities with promising results. Other problems arise not only from the performance of the PET detectors but also from the interference of the PET devices into the performance of the MRI system. Nonmagnetic components and materials must be employed to minimize these effects. The temperature variations inside the magnet can also affect the performance of the photodetectors and must be taken into account. The development of SiPMs, which have a higher gain than APDs, has opened new possibilities to improve this imaging modality [28].

10.3.2 Single-Photon Emission Computed Tomography

The intrinsic resolution and the efficiency of SPECT detectors can be improved using new scintillators. A gamma imaging detector with submillimeter intrinsic spatial resolution and 80% efficiency [2] has been developed employing a continuous LaBr$_3$ crystal and a PSPMT.

In SPECT, however, the performance is limited by the collimators that couple inversely the efficiency and the spatial resolution. Alternative techniques can bring substantial improvements. This is the case for Compton imaging, which applies the principle of Compton telescopes to medical imaging. Mechanical collimators are replaced by a first detector. Gamma rays Compton scatter in this detector and are absorbed in the second detector (Fig. 10.5). From the interaction position in the two detectors and the energy deposited in the first detector, it is possible to define a cone shell in which the emission point of the gamma rays is restricted. The intersection of the cones generated in several events defines the source location. The cone axis is given by the line connecting the two interaction points, the cone vertex is the interaction point in the first detector, and the angle of aperture is given by the Compton kinematics:

$$\cos\theta = 1 - m_0 c^2 \left(\frac{1}{E_0 - E_e} - \frac{1}{E_0} \right),$$

where E_0 is the energy of the incident photon, E_e is the energy of the recoil electron measured in the scatter detector, and $m_0 c^2$ is the electron rest mass.

The scatter detector must have excellent spatial resolution and low Doppler broadening. Doppler broadening is an uncertainty in energy deposited in the scat-

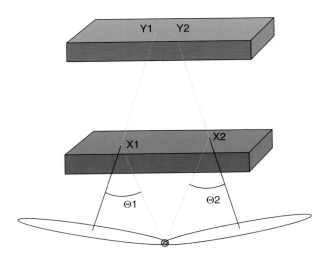

Fig. 10.5 Principle of Compton imaging

ter detector that occurs because the electrons involved in the Compton interaction are not at rest, producing a distribution of the energy deposited that depends on the material employed. Given its reduced Doppler broadening compared to other materials (only diamond has lower Doppler broadening), silicon is the optimum choice for the scatter detector. For the absorption detector, scintillator crystals or CZT or CdTe detectors are used.

Several groups are working in detector development for Compton imaging for different applications, for example, for a prostate probe imager, or small-animal PET and SPECT [29, 30]. A rat image has been obtained with this technique by employing a Compton detector with double-sided silicon detectors as scatterer and pixelized CdTe detectors as absorption detector. The distribution in the rat body of three radiotracers (In-111, I-131, and Cu-64) was imaged in vivo [31].

10.3.3 X-Ray and CT

Except for the amorphous selenium detectors for mammography, most X-ray detectors are based on indirect conversion of X-rays in CsI crystals and production of the signals in a photodiode, such as an amorphous silicon detector. The detectors integrate the electrical signals for a certain period, measuring the overall energy deposited. Research in this field aims at the development of direct conversion materials, which convert the energy from the absorbed X-rays directly into electrical signals; of photon-counting detectors to process the signals from each converted X-ray photon individually and of spectral detectors with energy-discriminating methods [32].

The use of direct conversion X-ray detectors can yield high spatial resolution and higher SNR, but they have problems of temporal artifacts due to charge trapping and, in some materials, dark currents due to the high bias voltage. Crystalline semiconductors would be the preferred solution, but due to the large areas to be covered with detectors with a small pixel size, amorphous or polycrystalline materials are used. Different materials are under investigation, such as amorphous selenium, GaAs, HgI_2, PbO, PbI_2, and CdTe or CZT in their crystalline or polycrystalline form. However, there is currently no direct conversion material that can replace scintillators in integrating detectors. Alternative photodetectors are also under investigation.

Spectral X-ray detectors need crystalline materials to achieve a fast enough response. GaAs or crystalline CdTe/CZT detectors can be used with the MEDIPIX chip, for example.

CT applications can considerably benefit from the use of photon-counting methods with multienergy discrimination. CZT and CdTe are the most promising direct conversion materials, but cost, performance stability, and mechanical robustness are still important issues.

References

1. C.W.E. van Eijk. Inorganic scintillators in medical imaging detectors. Nucl. Instr. Meth. A 509 (2003) 17–25.
2. R. Pani et al. Lanthanum scintillation crystals for gamma-ray imaging. Nucl. Instr. Meth. A 567 (2006) 294–297.
3. M.E. Daube-Witherspoon et al. The imaging performance of a LaBr3-based PET scanner. Phys. Med. Biol. 55 (2010) 45–64.
4. R. Mirzoyan, M. Laatiaoui, M. Teshima. Very high quantum efficiency PMTs with Bialkali photo-cathode. Nucl. Instr. Meth. A 567 (2006) 230–232.
5. W.W. Moses. Time of flight in PET revisited. IEEE Trans. Nucl. Sci. NS-50 (2003) 1325–1330.
6. R. Lecomte, R. A. deKemp and R. Klein. LabPET: a high-performance APD-based digital PET scanner for small animal imaging. J. Nucl. Med. 47 (2006) 194P.
7. V. Spanoudaki et al. Performance evaluation of MADPET-II, a small animal dual layer LSO-APD PET scanner with individual detector read out and depth of interaction information. J. Nucl. Med. 42 (S2) (2007) 39P.
8. P. Despres et al. Investigation of a continuous crystal PSAPD-based gamma camera. IEEE Trans. Nucl. Sci. 53 (3) (2006) 1643–1649.
9. B.J. Pichler et al. Performance test of an LSO-APD detector in a 7-T MRI scanner for simultaneous PET/MRI. J. Nucl. Med. 47 (2006) 639–647.
10. P. Buzhan et al. An advanced study of silicon photomultiplier. ICFA Instrum. Bull. 23 (2001) 28.
11. C. Piemonte. A new silicon photomultiplier structure for blue light detection. Nucl. Instr. Meth. A 568 (I 1) (2006) 224–232.
12. S-J. Park et al. A prototype of very high resolution small animal PET scanner using silicon pad detectors. Nucl. Instr. Meth. A 570(3) (2007) 543–555.
13. G. Di Domenico et al. SiliPET: an ultra-high resolution design of a small animal PET scanner based on stacks of double-sided silicon strip detector. Nucl. Instr. Meth. A 571 (1–2) (2007) 22–25.
14. P. Vaska et al. Ultra-high resolution PET: a CZT-based scanner for the mouse brain. J. Nucl. Med. 50 (S2) (2009) 293.
15. C.S. Levin, A.M.K. Foudray and F. Habte. Impact of high energy resolution detectors on the performance of a PET system dedicated to breast cancer imaging. Phys. Med. 21 (S1) (2006) 28–34.
16. G.L. Zeng and D. Gagnon. Image reconstruction algorithm for a spinning CZT SPECT camera with a parallel slat collimator and small pixels. Med. Phys. 31 (12) (2004) 3461–3473.
17. J. Missimer et al. Performance evaluation of the 16-module quad HIDAC small animal PET camera. Phys. Med. Biol. 49 (2004) 2069–2081.
18. M.-L. Gallin-Martel et al. A Liquid Xenon Positron Emission Tomograph for small animal imaging: First experimental results of a prototype cell. Nucl. Instr. Meth. A 599 (I2–3) (2009) 275–283.
19. E. Seravalli et al. 2D dosimetry in a proton beam with a scintillating GEM detector. Phys. Med. Biol. 54 (2009) 3755.
20. M.G. Bisogni et al. Performances of different digital mammography imaging systems: evaluation and comparison. Nucl. Instr. Meth. A. 546 (2005) 14–18.
21. A.P.H. Butler et al. Biomedical X-ray imaging with spectroscopic pixel detectors. Nucl. Instr. Meth. A 591 (I1) (2008) 141–146.
22. G. Llosá et al. Evaluation of the first silicon photomultiplier matrices for a small animal PET scanner. IEEE NSS-MIC 2008 Conf. Rec., M02-1. pp 3574–3580.
23. D. Schaart. A novel, SiPM based monolithic scintillator detector for PET. Phys Med. Biol. 54 (2009) 3501.
24. P. Bruyndonckx et al. Evaluation of machine learning algorithms for localization of photons in undivided scintillator blocks for PET detectors. IEEE Trans. Nucl. Sci. 55 (2008) 918–924.

25. T. K. Lewellen. Recent developments in PET detector technology. Phys. Med. Biol. 53 (2008) R287–R317.
26. S. Moehrs, et al. A detector head design for small-animal PET with silicon photomultipliers (SiPM). Phys. Med. Biol. 51 (2006) 1113–1127.
27. A.J. Lucas et al. Development of a combined microPET-MR system. Technol. Cancer Res. Treat. 5 (2006) 337–341.
28. Hyperimage simultaneous PET-MR imaging. http://www.hybrid-pet-mr.eu/
29. A. Studen et al. First coincidences in pre-clinical Compton camera prototype for medical imaging. Nucl. Instr. Meth. A 531 (II) (2004) 258–264.
30. G. Llosá et al. Last results of a first Compton probe demonstrator. IEEE Trans. Nucl. Sci. 55(3) (2008) 936–941.
31. N. Kawachi et al. Basic characteristics of a newly developed Si/CdTe Compton camera for medical imaging. IEEE 2008 NSS MIC Conf. Rec. NMR6. pp 1540–1546.
32. M. Overdick. Towards direct conversion detectors for medical imaging with X-rays. IEEE 2008 NSS MIC Conf. Rec. pp 1527–1535.
33. http://medipix.web.cern.ch/MEDIPIX/

Part V
New Frontiers in Nuclear Medicine

Chapter 11
The PET Magnifier Probe

Carlos Lacasta, Neal H. Clinthorne, and Gabriela Llosá

11.1 Introduction

Positron emission tomographic (PET) devices are routinely used for diagnosis. They provide medical doctors with a useful tool to locate and discover tumors with a spatial resolution that is in the order of 5–8 mm FWHM (full width at half maximum). This resolution is just enough for a search scan, but not enough to detect small lesions, making it challenging to detect uptake in tumors smaller than about 1 ml in volume, which translates to a 12.5-mm diameter of a spherical lesion. This is particularly true in the case of prostate and bladder cancers. Conventional PET scanning, with its large ring geometry, has not been effective for the detection of very small intraprostatic lesions, small pelvic lymph node metastases, invasion into nearby tissues, or in the case of bladder cancer, the extent of bladder wall invasion.

An increase of the spatial resolution in specific regions would be of great interest in situations like those mentioned as well as to monitor the developments of an already-known tumor for which one would like to contour the tumor, determine whether there is a single tumor or a cluster of small ones, and so on. This is what the concept of the PET magnifier probe intends to achieve. In this chapter, we survey the main ideas of the concept and how it can be implemented.

11.2 The PET Magnifying Probe Concept

To understand how the probe can increase the resolution of a PET device in certain regions of interest, we should remember how the image is obtained when we operate a PET device. PET devices work by detecting in coincidence a pair of photons

C. Lacasta (✉) and G. Llosá
Instituto de Física Corpuscular – IFIC/CSIC-UGEV, P.O. Box 22085, 46701 Valencia, Spain
e-mail: Carlos.Lacasta@ific.uv.es

N.H. Clinthorne
University of Michigan, Ann Arbor, MI 48109 USA

M.C. Cantone and C. Hoeschen (eds.), *Radiation Physics for Nuclear Medicine*,
DOI 10.1007/978-3-642-11327-7_11, © Springer-Verlag Berlin Heidelberg 2011

emitted, mainly back to back, by a positron generated in the radiotracer. The photons are often detected in a ring of scintillator detectors, which provide information about the impinging point and the energy. The two hit points define a line that we usually call line of response (LOR) and along which we believe the positron is annihilated into the two photons or, in other words, we have a point on the tumor volume. The crossing of many of these lines determines the tumor volume and position as sketched in Fig. 11.1 LORs are usually thought of as strips whose width is related to the scintillator intrinsic spatial resolution. Since all the crystals have the same resolution, the strip has uniform width all along the LOR.

Following this *geometrical* interpretation of the PET image reconstruction, it is not difficult to see that one way of increasing the overall spatial resolution of the PET instrument is to narrow the width of the LOR strip on one of its ends as shown in Fig. 11.2. Narrowing the LOR means, in this context, to have an instrument with better spatial resolution than the scintillators on that end of the LOR.

This is what the proposed PET magnifying probe is supposed to do. The probe, itself a PET detector, can be connected with present PET instruments at a point of modularity. It can be incorporated as an add-on detector to present and future PET devices. The use of small PET detectors having high resolution in conjunction with more conventional PET rings has been proposed in the past [1–3] for small animal and human imaging and has been examined in detail for high-resolution imaging of small animals [4, 5].

In the current proposal, the probe would be made of a number of stacks of 1-mm thick silicon pad (or big-pixel) sensors with a pad size of $1 \times 1\,mm^2$, providing as the minimum sensitive volume a $1\text{-}mm^3$ voxel with an intrinsic spatial resolution of about 1 mm FWHM.

The data sample used to reconstruct the image will contain two different sets of events. For the first, which we call low-resolution events, the LOR is made from two hits in the PET ring, and for the second set, the high-resolution events, the LOR is built from a hit in the probe and another one in the PET ring. The latter will have much better spatial resolution and, very likely, a much smaller occurrence rate. The way in which these two sets are treated by the reconstruction algorithm is not obvious and is subject to studies such that we can recover, somehow, the resolution of the probe even though the fraction of high-resolution events is much smaller than that of the PET ring alone.

11.2.1 The Gain in Resolution

Because the tomographic measurement process already entails significant uncertainty regarding where an annihilation occurs along each LOR, spatial resolution is primarily determined by uncertainty in directions transverse to the LOR. The component of uncertainty due to detector resolution at any point along the LOR in the transverse direction is approximately

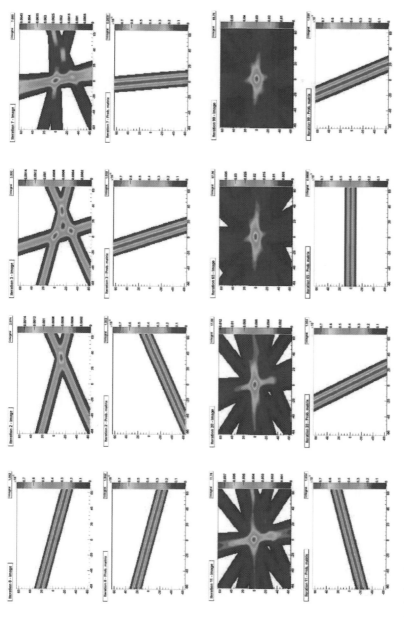

Fig. 11.1 A pictorial sketch of how the image is formed from the reconstructed lines of response. The accumulation of lines of response shows how their intersection provides the information about the position of a point source on the plane. Lines of response are *thick*, representing our degree of uncertainty. *Red* corresponds to a high probability of finding the source on that position and *blue* to a small probability

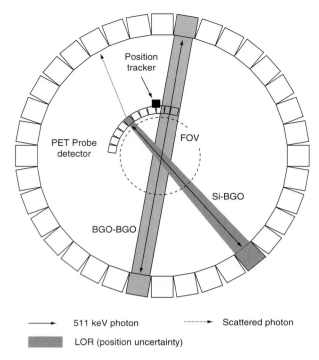

Fig. 11.2 Diagram of the positron emission tomographic (PET) magnifying probe concept. It shows pictorially the LOR made from scintillator (BGO) hits only and from the combination of silicon and scintillator (Si-BGO in the figure) hits. Adding another sensor with better spatial resolution will effectively narrow the line of response (LOR) in one of its ends, providing a better system spatial resolution in the region covered by the probe field of view

$$R_D \approx 2.35\sqrt{(1-\alpha)^2\left(\sigma_{D1}^2 \sin^2\theta_1 + \sigma_{C1}^2 \cos^2\theta_1\right) + \alpha^2\left(\sigma_{D2}^2 \sin^2\theta_2 + \sigma_{C2}^2 \cos^2\theta_2\right)} \text{ FWHM}$$

where θ_i is the angle of the LOR incidence in detector i; σ_{Di} and σ_{Ci} quantify the RMS (root mean square) uncertainty for each detector in the *depth* and *circumferential* directions, respectively; and α is the fractional distance along the LOR at which the uncertainty is desired. For example, if the distance from the probe to a source is 3 cm and the total LOR length is 40 cm, then $\alpha = 3/40$. Note that when α is small, spatial resolution is dominated by the characteristics of detector 1, which is the probe.

Figure 11.3 shows the resolution (quantified by LOR uncertainty) as a function of distance from the probe for an 80-cm diameter external ring having nominal resolution of 6 mm FWHM and for internal probe resolutions of 1, 2, and 3 mm FWHM. Below about 5 cm, the resolution is clearly dominated by that of the probe, and it can still be excellent even for 5–10 cm in the case of a 1-mm resolution probe and even further (up to 15 cm) for more modest resolutions of the probe.

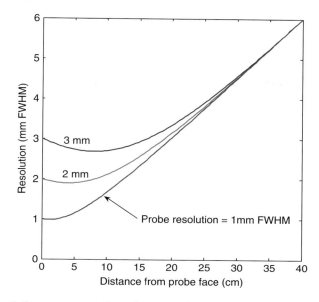

Fig. 11.3 Resolution transverse to lines of response (LORs) for coincidences between probes of 1-, 2-, and 3-mm resolution and an 80-cm diameter external ring with a resolution of 6 mm FWHM (full width at half maximum). For small distances, the overall resolution is governed by the probe resolution. This can be understood by the fact that at small distances of the probe we are in the region in which the LOR is narrower

11.2.2 Instrumenting the Probe

To build such a high-resolution probe, one needs to consider carefully the sensor material and its readout electronics. There are a number of issues that need to be studied. The material should be easily segmented to provide the required resolution, and it is also important that there is a trade-off between the resolution and the detector granularity since this will determine the number of electronics channels. Detection efficiency is of overriding importance in addition to the need for a material able to produce a fast enough signal to operate the probe in coincidence with the PET ring. A slow detector will require a too wide coincidence window with the corresponding increase of accidentals, resulting in an additional burden not only for the data acquisition system but also for the image reconstruction algorithms. However, this can be balanced by good energy resolution, which would allow rejection of those accidentals. Well-established sensor and readout electronics technologies are welcomed since that helps in narrowing the origin of any possible operational problem. Packaging turns out to be an important issue, not only because of the need for an apparatus that can be handled easily and therefore has to be compact but also because of the dependence of the resolution on the distance from the probe to the source. This forces the probe to be able to determine that distance, which needs segmentation in that direction and a relatively smooth variation of the resolution

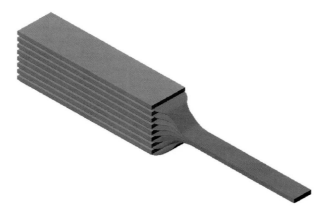

Fig. 11.4 Sketch of a stack of silicon sensors that will make the probe. The sensors and the cables that take the signal out to the readout electronics can be distinguished

along the probe thickness. We believe that silicon is a material that fulfills all those requirements, and this is why we are proposing to build the probe with stacks of 1-mm thick silicon pad detectors with $1 \times 1 \mathrm{mm}^2$ pads (see Fig. 11.4).

11.3 A Toy Simulation of the Probe

We have used a toy Monte Carlo simulation based on Geant4 [6] to illustrate the principles described. Figure 11.5 shows the basic geometry that we have considered. It consists of a 90-cm diameter PET ring that has a width of 20 cm. The ring intrinsic resolution in Φ is 4 mm. The *phantom* is an array of point sources with a separation of 7 mm in both directions. Figure 11.6 shows how the probe is implemented in the geometry. It consists of ten layers of $8 \times 8 \mathrm{cm}^2$ silicon pad sensors. The sensor thickness is 1 mm and the pad size is $1 \times 1 \mathrm{mm}^2$, providing a resolution of 1 mm. The gap between the silicon layers is 1 mm, and the distance from the probe to the source plane is 3 cm. From the simulation, the efficiency of low-resolution data is about 30% while the high-resolution data, with hits in the probe, is about 2%, depending on the distance from the probe to the source.

The reconstruction was done using a two-dimensional (2D) maximum likelihood-expectation maximization (ML-EM) algorithm. The Siddon [7] algorithm was used to calculate the system matrix. The top picture of Fig. 11.7 shows the image of the source array reconstructed with only the PET data. One can see some structure but with a considerable amount of blur. If we only use the probe data, we get the middle image of Fig. 11.7, in which the high resolution of the probe allows separation of the point sources along the x-axis, but the separation along the y-axis (which is the radial direction) is rather poor due to the small field of view of the probe alone. When we combine both data sets, we obtain the bottom image at Fig 11.7. There, we can really see the advantage of combining both data sets with different resolutions.

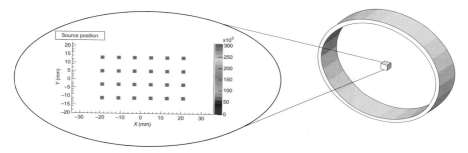

Fig. 11.5 Simulation setup used in Geant4 showing the array of sources located at the center of the ring. This corresponds to the image we want to reconstruct. The ring circumference is in the xy plane and so are the point sources

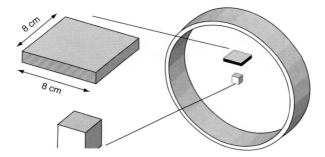

Fig. 11.6 Diagram of the geometry used in the fast simulation. We can see the source, which is an array of points 7 mm from each other; the probe, which is a stack of ten $8 \times 8 \times 0.1\,\mathrm{cm}^3$ together with the positron emission tomographic (PET) ring with a radius of 45 cm

We maintain the high resolution on the x-axis, while on the y-axis we can now separate clearly the source points. One can see, however, that there are some artifacts on the layer close to the probe, meaning that some mechanism should be implemented to weight properly the information in the different data sets.

11.4   Image Reconstruction

While filtered backprojection can be used to reconstruct images from the individual data sets, it cannot conveniently account for the varying resolutions and noise. Direct combination of both data sets results in domination by the low-resolution interactions, while weighting each inversely with its efficiency results in domination of the noise in the highest-resolution, lowest-efficiency data. A more suitable treatment of the issue is to embed the reconstruction problem within the framework of maximum likelihood estimation, which naturally accounts for the differing measurement quality (see chapter 12). As in conventional PET, we assume that the measurements y_i are Poisson distributed conditioned on the object distribution x through the following model [8]:

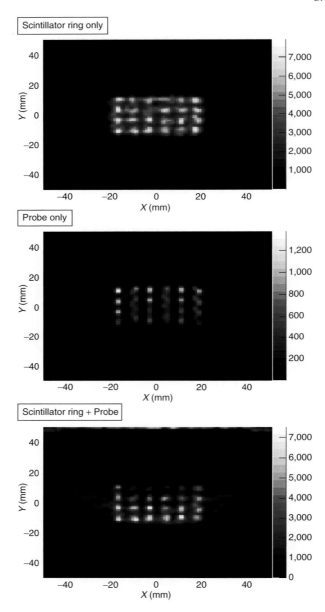

Fig. 11.7 Reconstruction of the array of sources in the simulation. *Top:* The image from the positron emission tomographic (PET) data alone. *Middle:* Image with the probe data alone. *Bottom:* Reconstruction with both types of events. The probe is located at $y = 30$ mm

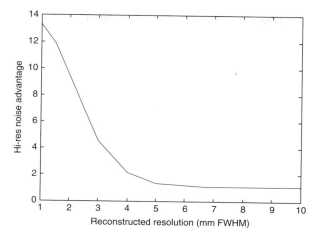

Fig. 11.8 Noise advantage of positron emission tomography (PET) with about 10% 1-mm intrinsic resolution data added to 4-mm resolution data compared to 4-mm resolution data alone. The curve is the ratio of standard deviations at a central point in reconstructed images for a large disk source. *FWHM* full width at half maximum

$$\begin{bmatrix} y_I \\ y_{II} \end{bmatrix} \approx Poisson \left\{ \begin{bmatrix} A_I \\ A_{II} \end{bmatrix} x + r + s \right\}$$

where A_i are the system matrices, and r and s are vectors representing the random coincidence and scatter, respectively. Note that to correctly weight the data, the different efficiencies and resolutions should be accounted for in the system matrices.

11.5 Expected Performance

From the simulation, we see that, on the whole data sample, the high-resolution data correspond to only a 10% of the total amount of data. The key question is whether such a small fraction of high-resolution information will significantly improve images. The results from the toy simulation seem to indicate so. If we assume about 10% high-resolution information in our simulation example, by using methods described in [8, 9], noise in the reconstructed images at any desired spatial resolution can be quantified and used as a reference for performance. Results of that calculation are presented in Fig. 11.8, in which the *ratio* of the standard deviations for a point at the center of the reconstructed images for the two systems is plotted as a function of desired spatial resolution in the reconstruction. The relative advantage of the small amount of high-resolution information is considerable when the desired spatial resolution is better than the intrinsic 4-mm resolution of the external ring. The performance, even with the addition of just a small amount of high-resolution data, increases rapidly.

11.6 The PET Magnifying Probe Design

Building a device suitable for medical imaging is a bit more than just testing a few sensors and seeking for cutting edge technologies that will solve the ultimate problems inherent in the field. This is certainly work that has to be done in early stages of the project. However, putting together all the components that make the device is also an important task. In particular, when we want it to perform as we believe it can. This is also the case for the PET magnifying probe, for which a number of different components have to be put together to build a reliable system. The different building blocks of the probe are mainly silicon sensors, readout electronics, connectivity of the former and the latter, a data acquisition system that should cope with the data stream at the given rates, and finally the appropriate image reconstruction algorithms that will allow full exploitation of the capabilities of the system. In this section, we discuss the most demanding components required for probe construction.

11.6.1 The Sensors

The sensors chosen for the PET probe are silicon pad detectors. The sensor is 1 mm thick, with 26×40 pads of 1×1 mm^2 each. The sensors have a p$^+$-doped implant, the pad itself, on an n-type substrate [10, 11]. To allow for easy wire bonding, the detectors have an extra metal layer on top of the implants where aluminum traces connect the implant to bonding pads on the edge of the sensor. Figure 11.9 shows a detail of the detector design showing those traces.

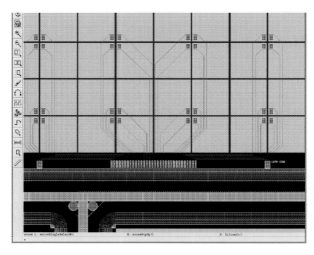

Fig. 11.9 Detail of the sensor design showing how the silicon pads (or big pixels) are connected via an aluminum trace to the bonding pads on the bottom side

11.6.2   Readout Electronics and Data Acquisition

The signal produced by the impinging photons in the silicon sensor has to be read out and digitized for further analysis and storage. For that, we have chosen an existing chip, the VATA GP7 [12]. The chip has a self-trigger signal that will be issued whenever 1 of its 128 channels registers a signal larger than a preprogrammed threshold. To accomplish that, the incoming signal is split into two branches. One goes through a fast shaper, 50-ns peaking time, and then to a comparator that will issue the trigger signal for values above the threshold. The second branch goes to a slow shaper, 500-ns peaking time, to accurately measure the signal amplitude.

To read out the silicon sensor with 1,040 pads, we need eight of these chips. The chips are mounted on an electronic board, the hybrid, that will provide the required voltages and control signals to the chips as well as the bias voltage for the sensors.

The data are transferred to a computer using a VME[1] bus. The bit streams used to configure the chip behavior properly are also sent through the VME bus. A data acquisition program [13] that controls the configuration and data transfer will run on the computer. The program will also monitor the data quality and save the data on disk for further analysis.

11.6.2.1   Timing

The probe has to work in coincidence with a scintillator ring. That makes timing an important issue, first to associate properly the hits in both detectors and second to reject as many random coincidences as possible. The former is of chief importance, in particular when high rates are expected, while the latter is an issue that may affect efficiency if the data acquisition system is not smart enough to make the rejection either online or off line by applying some cinematic cuts.

The coincidence will be built by looking at the scintillator ring trigger signal, which is assumed to be fast, and at the VATA GP7 self-trigger signal, which is not that fast. This, per se, would not be a problem since we could do with a fixed delay between the two systems. However, this is not the case. The signal formation in a 1-mm thick slab of silicon can take as long as about 100 ns, depending on where, along the detector thickness, the photon deposits its energy. This position is somewhat random following an exponential distribution peaking at the entry surface of the detector. This will introduce a spread of the signal formation time in the silicon. To that, we should add the time walk associated with the signal discrimination at the comparator. The time spread associated with the time walk will depend on the increase time of the signal, 50 ns in our case, and on the signal-over-threshold ratio. The closer the signal is to the threshold, the bigger the delay will be. Since we

---

[1] VME is a computer bus standard which has been widely used for many applications, in particular in Particle Physics experiments, and standarized by the IEC (International Electrotechnical Commission) as ANSI/IEEE 1014-1987.

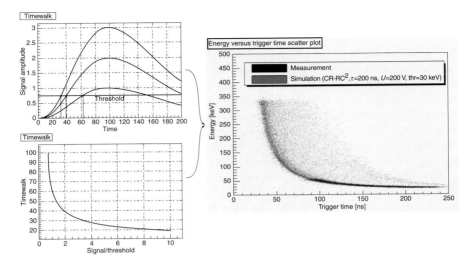

Fig. 11.10 For a given signal increase time, the time at which the amplitude equals the threshold depends on the signal-over-threshold ratio. This effect is called time walk, and the concept is shown pictorially at the *upper left*. The dependence of the delay on the signal-over-threshold ratio is shown in the *lower left*. The figure at the *right* shows the comparison of the measured data and a simulation. In the latter, we included the spread of the signal formation time in silicon convoluted with the proper energy spectrum in silicon

are measuring Compton interactions, the energy is not fixed; therefore, we end up with a time spread due to time walk. Figure 11.10 shows the concept together with a comparison of measured data with a full simulation of the experimental setup [14].

There a number of knobs we can play with to optimize the timing performance when designing the probe. On one side, we can increase the electric field in the sensitive area of the silicon detectors so that charges will drift faster, accordingly reducing the signal formation time in the silicon. On the other hand, changing the electric field shape can also help. In fact, increasing the pad size produces more uniform fields and somewhat faster signals [15]. However, this is at the expense of the achievable spatial resolution. Finally, we can have faster electronics to reduce the effect of the time walk; again, this will increase the noise and the power consumption of the chips, which will translate to an increase of the system temperature, which can be an issue if it rises too high.

11.6.3 Prototypes

We have tested thoroughly the existing individual components of the probe, mainly the sensors and the chips. We have assembled a module composed of a silicon 1,040 pad sensor and eight of the readout chips as shown in Fig. 11.11; this has performed as expected from the results of the individual components. The next step is the design of a stack of modules. For this prototype, the idea is to place a double-sided, eight-chip hybrid at each side of a structure made by two back-to-back sensors,

Fig. 11.11 *Top*: One of the prototype modules for the probe. The silicon pad detector is seen on the center with two single-sided, four-chip hybrids on each side. *Bottom*: Mechanical three-dimensional (3D) computer-aided drawing (CAD) model of the probe prototype. This prototype will be built with four modules. Each module would be made by two sensors glued back to back read out by a double-sided, eight-chip hybrid on each side

which would make a *module*. The probe prototype would consist of a stack of such modules, as shown in the three-dimensional (3D) computer-aided drawing (CAD) mechanical model at the bottom of Fig. 11.11.

11.6.4 Packaging

As shown in Fig. 11.3, the final system resolution depends on the distance from the probe to the source. This dependency is almost linear with a nonnegligible slope at distances above a few centimeters. To avoid large variations of the resolution along the probe thickness, it is of paramount importance to achieve the highest packaging factor possible; that is, the fraction of thickness not due to sensitive volume should be reduced to the minimum.

The packaging factor is usually driven by the interconnection among the different electronic components, in particular the sensor and the readout chips. To address this important aspect of the device, a microcable technology, similar to TAB (tape-automated bonding), has been explored [16, 17]. This technology consists of

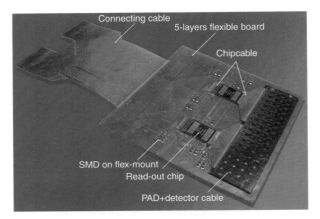

Fig. 11.12 View of a prototype module using the microcable technology. Even the passive components in the hybrid are tape-automated bonded (TAB), allowing for a huge packing factor

connecting, using a special ultrasonic bond wedge, the pads of an object to aluminum traces printed on a polyimide flex, the cables, and carrying the electronic signals to the next connection pads. This technology has some advantages with respect to the industrial high-volume TAB process. In particular, it can manage much smaller pitches and thinner and more flexible flexes. The current design can achieve 47-μm pitch and a thickness of 10 μm. However, to reduce the trace capacitance, we have added a dielectric insert that is 40 μm thick. All in all, the overall cable thickness is about 50 μm. This allows a packing factor higher than 90%. Figure 11.12 shows a prototype of a module built with this microcable technology. One can see that there are different cables, one per chip, another one for the sensor, one for the hybrid, and finally the connection cable. This allows a modular progress by which one can use conventional wire bonding for some parts of the module while testing the different microcables.

11.7 Summary

We have reviewed the concept of a PET magnifying probe to be used as an add-on to existing or future PET devices to increase the spatial resolution in some areas of interest. Although this is still an ongoing research project, it has proven to be a promising device. We have seen how the probe resolution can be recovered, even if the fraction of high-resolution events is much lower than that of the worse resolution events. Results from a toy simulation have also shown the increase of resolution but using a simplistic reconstruction algorithm. Effort is being invested in reconstruction algorithms that will combine the two kinds of events properly.

The probe is realized as a number of stacks of silicon pad sensors providing an intrinsic spatial resolution of about 1 mm FWHM. Prototype modules, the assembly of sensor plus readout chips, have been built showing the expected performance and we are in the process of building the first probe prototype by stacking several of those modules.

References

1. Clinthorne NH, Meier D, Hua C, Weilhammer P, Nygard E, Wilderman SJ and Rogers WL. *Very high resolution animal PET.* J. Nucl. Med., vol. 41, p.21, 2000.
2. Clinthorne NH, Park S, Wilderman SJ, Rogers WL. *High resolution PET detector.* J. Nucl. Med., vol. 42, p. 102, 2001.
3. Park S, Rogers WL, Wilderman SJ, Sukovic P, Han L, Meier D, Nygard E, Weilhammer P and Clinthorne NH. *Design of a very high resolution animal PET.* J. Nucl. Med., vol. 42, p. 55, 2001.
4. Park SJ, Mikuz M, Lacasta C, Kagan H, Weilhammer P, et al. Performance evaluation of a very high resolution small animal PET imager using silicon scatter detectors. Physics in Medicine and Biology 52: 2807–2826, 2007.
5. Park SJ, Rogers WL, Clinthorne NH. Design of a very high-resolution small animal PET scanner using a silicon scatter detector insert. Physics in Medicine and Biology 52: 4653–4677, 2007.
6. Agostinelli S et al. Geant4 – a simulation toolkit. Nuclear Instruments and Methods A 506: 250–303, 2003.
7. Siddon RL. Fast calculation of the exact radiological path for a three-dimensional CT array. Medical Physics 12: 252–255, 1985.
8. Clinthorne NH, Sangjune P, Rogers WL, Ping-Chun Chiao. Nuclear Science Symposium Conference Record, 2003 IEEE. Vol. 3, 1997–2001.
9. Meng LJ, Clinthorne NH. *Modified uniform Cramer–Rao bound for multiple pinhole aperture design.* IEEE Trans. On Nuclear Science, Vol. 23, No. 7, 896–902, July 2004.
10. Meier D et al. Silicon detector for a Compton camera in nuclear medical imaging. IEEE Transactions on Nuclear Science 49: 812–816, 2002.
11. Studen A et al. Development of silicon pad detectors and readout electronics for a Compton camera. Nuclear Instruments and Methods A 501: 273–279, 2003.
12. VATAGP7 + Gamma Medica – IDEAS reference.
13. Lacasta C. DAQ++: A C++ Data Acquisition Software Framework. IEEE Real-Time Conference Record, 2007.
14. Studen A et al. Timing in thick silicon detectors – an update. Nuclear Instruments and Methods in Research Section A, Accelerators, Spectrometers, Detectors and Associated Equipment 579(1): 83–87, 2007.
15. Mikuž M, Studen A, Cindro V, Kramberger G. Timing in thick silicon detectors for a Compton camera. IEEE Transactions on Nuclear Science 49: 2549–2557, 2002.
16. Stankova V et al. Development and test of micro-cables for thin silicon detector modules in a prostate probe. Nuclear Science Symposium Conference Record, 2008. NSS '08. IEEE 19–25 Oct. 2008: 4571–4574.
17. Lacasta C et al. Development and test of TAB bonded micro-cables for silicon detectors in a Compton prostate probe. Nuclear Science Symposium Conference Record, 2005 IEEE Volume 5, 23–29 Oct. 2005: 3032–3035.

Chapter 12
Algorithms for Image Reconstruction

Christoph Hoeschen, Magdalena Rafecas, and Timo Aspelmeier

12.1 Introduction

Three-dimensional (3D) imaging is becoming one of the most important applications of radioactive materials in medicine. It offers good spatial resolution, a 3D insight into the human body, and a high sensitivity in the picomolar range because markers for biological processes can be detected well when labeled with radioactive materials. In addition, the technical equipment has undergone many technological achievements. This is true for single-photon emission computed tomography (SPECT), positron emission tomography (PET), and X-ray computed tomography (CT), which is often used in connection with the nuclear medical imaging systems, as also described in chapter 5 about sources in nuclear medicine. As can be realized by the names of the systems, the imaging methodologies all generate the images using a computational process. This is necessary since in all types of CT the purpose is to generate a stack of two-dimensional slices (a 3D data set) that are reconstructed from various "projections" along certain lines. This reconstruction process can be achieved by various different methods, which can be divided into so-called algebraic or iterative reconstruction methods and analytical methods. After a brief introduction to give an approach to the reconstruction task in general, we describe both kinds of algorithms.

C. Hoeschen (✉)
Helmholtz Zentrum München, German Research Center for Environmental Health GmbH, Ingolstädter Landtstr. 1, 85764 Neuherberg, Germany

M. Rafecas
IFIC Instituto de Física Corpuscula CSIC-Universitat de València, Edificio Institutos, de Investigación Ap. Correos 22085, 46071 Valencia, Spain

T. Aspelmeier
Scivis wissenschaftliche Bildverarbeitung GmbH, Bertha-von-Suttner-Str. 5, 37085 Göttingen Germany

M.C. Cantone and C. Hoeschen (eds.), *Radiation Physics for Nuclear Medicine*,
DOI 10.1007/978-3-642-11327-7_12, © Springer-Verlag Berlin Heidelberg 2011

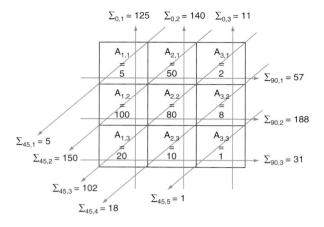

Fig. 12.1 A 3×3 twodimensional matrix of an activity distribution in the object to be investigated and various detection directions from which the original distribution has to be determined

12.1.1 Mutual Understanding

As a first approach to the mathematical and computational procedures, we explain what is the basic task of such a reconstruction process. For that, we assume a two-dimensional matrix of 3×3 elements representing, for example, for nuclear medical imaging, a distribution of activity in an investigated object. We assume measurement points in various directions, resulting in a situation as presented in Fig 12.1.

As we see in the figure, the data we measure are not the information from which we can describe the object or, especially in the case investigated here, the activity distribution directly; we do need to reconstruct the activity values in our object. This can be done by describing the measured values and their corresponding activity sums in a set of linear equations in the form

$$\sum_i A_i = A_{\text{tot,meas}} \tag{12.1}$$

assuming that the emitted radiation is not absorbed, which is certainly not the case. This is one of the most important reasons for performing X-ray CT investigations in connection with nuclear medical imaging approaches since the CT information can be used for absorption correction.

If we solve such a set of linear equations, which is possible if there is a sufficient number of independent equations (meaning the set of equations is orthogonal regarding the number of activities to be determined), we get the activity distribution matrix desired. However, if the matrix becomes larger, the amount of equations rises quickly; therefore, the computation time becomes problematic. In addition, due to the limited amount of activity that can be applied to patients, the measured activity sums per direction are noisy values, meaning that the set of equations might not be

solvable if measurements are not done at exactly the same time point and assuming
a completely uniform radiation distribution. This is in general not the case, which
means that solving a set of linear equations is not the solution that is suitable for better resolved image slices and noisy data. Before we show what can be done instead
to solve the problem of reconstructing an activity distribution from measurement
values at the outside of an object, we should mention that the problem in X-ray CT
is similar. In that case, we do measure the line integrals over a variety of averaged
absorption coefficients within the object elements. This would again result in a set
of linear equations, at least as long as we assume monoenergetic X-rays used for the
imaging process. In that case, we would have

$$I = I_0 \cdot e^{-\sum_i \mu_i} \tag{12.2}$$

resulting in

$$\sum_i \mu_i = -\ln\left(\frac{I}{I_0}\right) \tag{12.3}$$

thus describing a similar task as the formula of Eq. 12.1 for the 3D nuclear medical
imaging task. The results discussed here are therefore in principle valid for both
types of imaging modalities.

12.2 Reconstruction with Analytical Methods

Let us assume that the object, which might be the intensity distribution of the
radiopharmaceutical or the distribution of the absorption coefficients in X-ray-based
tomography, is represented by $f(x, y)$. This object function is what has to be reconstructed. With $m(d, \theta)$, we describe the measurements or projections measured from
or assumed to be parallel rays to the detectors aligned in angle θ; d is the distance of
the ray of the projection from the isocenter. With this notation, we can formulate the
basis of the most used method for reconstructing data of a slice from its projections
or projectional measurements analytically as defined by the Austrian mathematician
Radon in 1917 [1] as discussed next.

12.2.1 Fourier Slice Theorem Principle

The Fourier transform of a parallel projection of an object described by $f(x, y)$
obtained at an angle θ equals a line in the two-dimensional Fourier transform of
such $f(x, y)$ taken at the same angle θ.

Various possibilities exist to prove such theorem. We briefly describe a method
using a rotated coordinate system. The rotation is done so that one axis is parallel
to the path of the projections with angle θ. We notate this axis by s. We can than
rewrite the description of the object by $f^*(d, s)$ using the following equations for
the coordinates:

$$d = x \cos \theta + y \sin \theta$$
$$s = -x \sin \theta + y \cos \theta$$

By integrating $f^*(d,s)$ along the s axis, we achieve the projection measurement $m(d,\theta)$ as

$$m(d,\theta) = \int_{-\infty}^{+\infty} f^*(d,s) \mathrm{d}s. \tag{12.4}$$

The Fourier transform over variable d yields

$$M(R,\theta) = \int_{-\infty}^{+\infty} \int_{-\infty}^{+\infty} f^*(d,s) \mathrm{d}s \, e^{-i^2 \pi R t} \mathrm{d}\,d. \tag{12.5}$$

Using coordinate transformation and the theory of calculus regarding coordinate transformation and differentiation, we achieve

$$M(R,\theta) = \int_{-\infty}^{+\infty} \int_{-\infty}^{+\infty} f(x,y) e^{-i2\pi R(x \cos \theta + y \sin \theta)} \mathrm{d}x \, \mathrm{d}y. \tag{12.6}$$

Using

$$u = R \cos \theta \quad \text{and} \quad v = R \sin \theta,$$

we get

$$F(u,v) = F(R \cos \theta, R \sin \theta) = M(R,\theta),$$

which describes the Fourier slice theorem [2]. It means that we can fill the entire Fourier space of the object description if we collect enough projections in the range between 0 and θ and can therefore reconstruct the object description from such projections. However, there are a number of problems in such a theoretical scheme of reconstructing a slice from the corresponding projections. With the projection data collected, typically there is no equidistant filling of the angular space as required according to the formulas mentioned. This means that interpolation is required, which will reduce the possible resolution in the reconstructed slice.

12.2.2 Backprojection

In addition, if we backproject every measurement value to the whole ray, a single source point would result in a starlike formation around that point in the reconstructed image, which could be referred to as an artifact. This is because there are no negative components in the reconstruction scheme with simple backprojection [3]. Therefore, it is necessary to implement a filter in the process to improve the edges and by that introduce negative values, which allow an artifact-free reconstruction of the slices. In theory, a weighting of the information with the frequency of the detail would allow a perfect reconstruction as described in the following formula which is the filtered backprojection (FBP) [4]:

$$f(x, y) = \int\limits_{0}^{\pi} d\theta \int\limits_{-\infty}^{+\infty} M(\omega, \theta) |\omega| \cdot e^{j 2\pi \omega (x \cos \theta + y \sin \theta)} d\omega \qquad (12.7)$$

This reconstruction process is directly achievable from the Fourier slice theorem as stated. With that filtering process, the noise is also increased. So, as a matter of optimization it is necessary to optimize the filter for the imaging task. In modern CT systems as well as for nuclear medical imaging devices, there is often a variety of filters to choose from depending on the specific imaging task. To use this reconstruction scheme for real measurement data, it is necessary to discretize the reconstruction formula. Doing so, one obtains

$$f(x, y) \approx \sum_{j=0}^{N-1} \int\limits_{-\infty}^{\infty} R_f(\theta_f, s) h(t - s) ds \qquad (12.8)$$

where R_f denotes the measured value for the activity sum or the absorption sum for a discrete distance s from the central axis and a discrete angle of detection; h is the discretized filter function.

12.2.3 Orthogonal Polynomial Expansion on the Disk

A completely different scheme for reconstruction [5] uses the fact that objects limited to a certain area and under certain conditions can be approximated by an orthogonal polynomial expansion.

We assume f represents the image of the object as a function on the unit disk. Then, V_n denotes the space of orthogonal polynomials on B^2 of total degree n as described by

$$\int_{B^2} P(x, y) Q(x, y) dx \, dy = 0 \quad \text{with} \quad \deg Q < \deg P = n \qquad (12.9)$$

Then, the following holds according to the first sentence in this paragraph:

$$L^2(B^2) = \sum_{k=0}^{\infty} \oplus v_k \quad \text{and} \quad f(x) = \sum_{k=0}^{\infty} \text{proj}_k f(x) \qquad (12.10)$$

The infinite series holds in the sense that the sum

$$S_n f(x) = \sum_{k=0}^{n} \mathrm{proj}_k f(x) \tag{12.11}$$

converges to f if $n \to \infty$. The polynomial $S_n f$ provides an approximation to the function $f \cdot S_n f(x)$ converges to $f(x)$ uniformly on B^2 if f has continuous first-order derivatives.

Using suitable polynomials, one can write this as

$$f \sim S_N(f; x, y) = \frac{1}{\pi} \sum_{v=0}^{N} \int_{-1}^{1} \Re_f(\phi_v, t) \Phi_v(t; x, y) \mathrm{d}t \tag{12.12}$$

with Φ_v polynomials of two variables given by exact formulas. If one finds an appropriate quadrature formula, one can obtain the approximation of the object function directly from a sum of the Radon data given by the measurement procedure multiplied by the Tschebychev polynomials as described in the following representation:

$$A_N(f; x, y) = \sum_{v=0}^{N-1} \sum_{j=0}^{N-1} \Re_{\phi_v}(f; \cos \psi_j) T_{j,v}(x, y) \tag{12.13}$$

Note that the data needed for this reconstruction procedure are not distributed uniformly along the axis but equally in angles around the disk for which the orthogonal polynomial expansion is performed. Various configurations can be assumed for different numbers of measurement points as well as different resolutions of the images and 3D problems as well. One advantage of this methodology compared to filtered backprojection is that no filtering is needed, resulting in better signal-to-noise ratios, especially for higher-frequency details and more accurate reconstructed data for a correct representation of the object.

Both methodologies will have measurement configurations for which they might fit better than the other analytical method and might therefore outperform the other one. It is important to find the better-fitting one for the task in question to obtain the most satisfactory results possible for the lowest radiation dose achievable.

12.3 Iterative Reconstruction Methods

12.3.1 Introduction

In the preceding section, the basics of analytical reconstruction techniques were described. One of the main drawbacks of analytical reconstruction methods is that they rely on a rather ideal model of the imaging process. This model assumes that

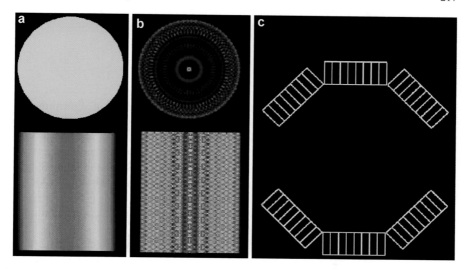

Fig. 12.2 Effect of inhomogeneous sampling: (**a**) the ideal sinogram (*bottom*) of an ideal homogeneously filled disk (*top*); (**b**, *bottom*) the sinogram resulting from binning the data from an homogeneously filled cylinder acquired using the positron emission tomographic (PET) scanner prototype MADPET-I in 20 rotation steps; (**b**, *top*) the corresponding reconstructed image using filtered backprojection. (**c**) Schematic representation of the scanner MADPET-I

measured data can be described by line integrals of the object distribution [1], ignoring the size of the detectors as well as other physical effects involved in the detection of the radiation. Moreover, analytical methods require complete and homogeneous data sampling. However, missing data can arise, for instance, from detector gaps and may cause severe artifacts in the reconstructed image; alternative system configurations yield nonuniformly sampled data. (An example of how an inhomogeneous sampling can distort the reconstructed image when using analytical reconstruction techniques is shown in Fig. 12.2) In addition, statistical noise, which is inherent in any radiation measurement, is not considered in the model. Before reconstruction, data can be processed to compensate for some of these undesired effects. Alternatively, iterative reconstruction techniques can be used.

While the assumptions on which analytical methods are based usually hold in CT, this is often not the case in emission tomography (ET), both PET and SPECT. We thus focus on ET methods in the following.

12.3.2   Main Components of Iterative Reconstruction Algorithms

The principle of iterative reconstruction algorithms is to find a solution by successive estimates. In the literature, a wide variety of iterative reconstruction techniques has been presented. Further information is provided in some excellent reviews [6–13]. Most of these techniques, however, share the same "ingredients" [8, 14]:

1. Image model
2. Data model
3. System model
4. Objective function
5. Optimization algorithm

Obviously, the performance and characteristics of the reconstruction algorithm will depend on the choice for each of these components; however, the data model (statistical or not) is the basis of a general classification of iterative methods: On the one hand is the family of statistical reconstruction techniques; on the other hand are the algebraic reconstruction methods, which are nonstatistical. Since the latter ignore the issue of statistical noise in the data, algebraic reconstruction techniques (ARTs) are often better suited for CT than expectation–maximization (EM) [15].

The emphasis in this discussion now turns to statistical reconstruction techniques since the frame of the following section is ET.

12.3.3 Ingredient 1: Image Model

The unknown tridimensional radiotracer distribution can be described by a function $f(x, y, z)$. To represent the object as a discretized image, this function can be approximated by a linear combination of a finite number J of basis functions $b_i(x, y, z)$:

$$f(x, y, z) \approx \sum_{i=0}^{J-1} f_i b_i(x, y, z) \qquad (12.14)$$

where f_i are the coefficients of the linear expansion.

The most popular choice for the basis element is the "voxel," although other basis functions, such as "blobs" [16], "natural pixels" [17], or "polar pixels" [18] can be found in the literature. A vector f is then formed using the values of the coefficients f_i lexicographically ordered: $f^T = \{f_0, f_1, \ldots, f_{J-1}\}$; J is the number of voxels in the image. The image intensity is considered constant over the cubic support of the voxel and zero elsewhere. The goal of image reconstruction is then to find an estimation for vector f.

Statistical methods can be, again, classified into two main groups, depending on whether vector f is considered deterministic or random: In a *classical* scheme, we assume that data alone determine the final solution, while in a *Bayesian* framework some previous knowledge of the experimenter can be translated into the algorithm by considering that f is a random vector with a prescribed probability prob(f). This probability is called the *prior* and reflects the expectations of the experimenter; for example, we can assume that PET images are smooth. Then, a low probability is assigned to nonsmooth images. If images from another modality, such as CT or magnetic resonance (MR) are available, the information about organ boundaries contained in these images can be used to define priors for reconstruction of PET data so that smoothness is only enforced within the same boundary.

12.3.4 Ingredient 2: Data Model

The iterative reconstruction methods most commonly used in EM are based on prob-abilistic models to incorporate the statistical nature of the data within the algorithm. The measured data (which can be described by a vector g) are now considered a representation of a random vector, which is governed by a probability distribution.

The meaning of an element g_i of the measurement vector g depends on how the data are organized: Measured data are commonly processed and stored into sino-grams; in this case, the subindex i refers to the sinogram bin i, and g_i is the number of counts assigned to that bin. However, the binning process required to build a sinogram might lead to some loss of spatial resolution. In opposition to analytical reconstruction methods, iterative techniques are flexible and allow other forms of organizing the data.

Data can be also stored according to the line-of-response (LOR) index i, lex-icographically ordered. In that case, g_i represents the number of counts detected by LOR i. For pixelated detectors, this kind of histogramming preserves spatial resolution.

To describe the probability distribution of each measurement g around its expec-tation value $E[g]$, a Poisson model is usually adopted because photon detection and photon emission obey a Poisson distribution. The assumption that g_i are indepen-dent Poisson variables holds as long as no correction factors have been previously applied and detector dead time can be neglected. Otherwise, other models should be preferred (e.g., a shifted Poisson model for PET data corrected for randoms).

The model of choice determines the criterion to select the "best" estimate for the true image. In a statistical framework, the best solution is the most likely one.

12.3.4.1 Poisson Model

The Poisson model is the most commonly used model in ET. For an object described by f, the probability of measuring g_i counts is described by the Poisson law

$$\text{prob}(g_i; f) = \frac{\overline{g_i}^{g_i} e^{-\overline{g_i}}}{g_i!}, \tag{12.15}$$

where $\overline{g_i}$ is the ith element of $E[g]$.

We can assume that the elements g_i are independent random variables; the probability law for g is then

$$\text{prob}(g; f) = \prod_{i=0}^{I-1} \text{prob}(g_i; f). \tag{12.16}$$

12.3.4.2 Gaussian Model

Preprocessing of the data often results in probability distributions that deviate from the Poisson model. If the number of counts per LOR is high enough, the Gaussian probability density function represents a good approximation:

$$\text{prob}(g_i; f) = \frac{1}{\sqrt{2\pi}\sigma_j} \frac{e^{(g_i - \overline{g_i})^2}}{2\sigma_i^2}, \tag{12.17}$$

The term σ_i^2 represents the variance for each LOR i. In principle, the variance is unknown but can be approximated according to the preprocessing steps or simply setting the variance equal to the noisy counts $\sigma_i^2 = g_i$. Since the latter approach can lead to a negative bias in the reconstruction and, as a consequence, can cause image artifacts, σ_i^2 can also be estimated from a smoothed version of the data or from projecting the current reconstructed image.

12.3.5 Ingredient 3: System Model

The system model relates the unknown image f to the expectation of each measurement g_i. In principle, the expected number of counts detected in LOR i can be described by the following linear, spatially variant model:

$$\overline{g_i} = \int_{\text{FOV}} d\mathbf{r} f(\mathbf{r}) h_i(\mathbf{r}) \tag{12.18}$$

where $\mathbf{r} = (x, y, z)$, and $h_i(\mathbf{r})$ is the system response for LOR i to an emitting source located at $\mathbf{r}$; $h_i(\mathbf{r})$ is also called the *point spread function* (PSF) of the system. In theory, any physical linear effect can be modeled into the PSF, such as attenuation, scatter, detector geometry, crystal penetration, detector resolution, and so on.

As mentioned, iterative algorithms use a discretized version of the image, described by the vector f. Then, Eq. 12.18 can be approximated by a system of linear equations:

$$\overline{g_i} = \sum_{j=0}^{J-1} H_{ij} f_j \tag{12.19}$$

Now, H_{ij} is an element of the *system response matrix* $\mathbf{H}$; H_{ij} represents the probability of detecting an emitted photon (SPECT) or pair of photons (PET), originating in voxel j, by the LOR i. The system matrix $\mathbf{H}$ thus consists of $I \times J$ elements, and it can be large. (For instance, in 3D PET, the size can be on the order of 10^6–10^9.

Equation 12.19 represents a forward projection operation inherent in the image formation process, and it can be summarized by the single matrix equation

$$\overline{g} = Hf. \tag{12.20}$$

12.3.5.1 Inverse Problems

The reconstruction problem consists of solving Eq. 12.20. Reconstructing an object f from its projections g given a physical system described by $\mathbf{H}$ is a typical example of a *linear inverse problem*.

We consider that a problem is a direct problem when it follows the sequence "cause-effect": Given the causes, we have to calculate the corresponding effects according to some physical laws. In contrast, an *inverse problem* is associated with the reverse sequence: Our goal is finding the unknown causes (the radioactivity distribution) of known consequences (the measured data). In general, solving an inverse problem is much more difficult than solving a direct problem. One reason is that, in the sequence cause-effect some information loss occurs.

Expression 12.20 might suggest that the solution of the underlying inverse problem can be found by inverting the equation. Except for the simplest cases, when H is a well-behaved invertible matrix, there are three problems that can arise from Eq. 12.18. If one or more of these problems exist for a particular application, the inverse problem is called *ill posed*. First, the equation may have no exact solution; mathematically this means that g is not in the span of H. This can happen, for example, if the projections are noisy, which is always the case in practice. The projections are then likely to have a noise component that lies outside the span of H. Second, there may be more than one solution, which is the case when H is such that it maps a subspace (the kernel of H) to 0. Any component of f that lies in this subspace does not contribute to Hf and is thus arbitrary. These problems can even appear simultaneously; that is, even though there is no exact solution, a part of f is still arbitrary.

There is, however, the third problem which is the most difficult and which can occur even when H is formally invertible. Often, the matrix H is *ill conditioned*, which means that certain directions are suppressed (i.e. nearly, but not exactly, mapped to 0) by H, while others are enlarged. When this happens, any noise component in the suppressed directions that is present in g is amplified by the inversion and gives a reconstruction f that is far away from the true object sought.

In addition, the large size of H also hinders the use of the direct inversion techniques, and iterative procedures are used instead.

12.3.5.2 Computation of H

The large size of H makes it impossible to store all elements. When object scatter is not modeled in H, the system matrix is highly sparse. Even so, the high number of nonzero elements makes the forward and backprojection operations computer expensive. An approach to facilitate the calculation is to factorize H into several matrices [10]. In PET, H can be represented as

$$H \approx H_{\text{det.sens}}H_{\text{det.blur}}H_{\text{attn}}H_{\text{geom}}H_{\text{positron}}. \qquad (12.21)$$

The principal component is H_{geom}, which describes the geometrical mapping between the source and the data: $(H_{\text{geom}})_{ij}$ corresponds to the probability that a photon pair originating in voxel j reaches the front faces of the detector pair described by the LOR i (see Fig 12.3). The physics of the detection process are mainly modeled in $H_{\text{det.blur}}$, which acts as a local blurring function; $H_{\text{det.blur}}$ can include effects such

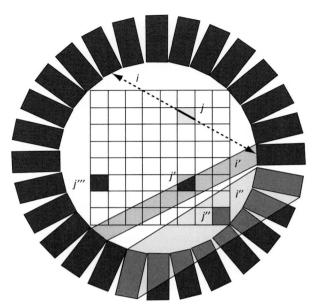

Fig. 12.3 Geometrical modeling of the system response. The simplest model assigns to H_{ij} the length of the intersection between the line of response (LOR) i (*dashed line*) and the voxel j; more accurate modeling takes into account the volume of response (VOR) i' defined by the faces of the detector crystals and assigns to $H_{i'j'}$ the intersection area between VOR i' and the voxel j'; however, the latter model neglects crystal penetration effects, which require consideration as VOR the volume i'', comprehended by the two detectors and to compute $H_{i''j''}$ taking into account the response function within that volume. In any case, the sparsity of the system matrix for each model can be easily seen: Only voxels intersecting the LOR i or the VORs i', i'' will contribute to the matrix. In all cases, the element related to voxel j'' will be zero

as photon accollinearity, scatter between detectors, or crystal penetration. However, the variations in the intrinsic sensitivity of individual crystals are modeled separately in $H_{\text{det.sens}}$. While all these components only depend on the scanner, H_{attn} and H_{positron} are object dependent. Attenuation in the patient is modeled in H_{attn}, which contains the survival probabilities of the photons. The effect of positron range is modeled into H_{positron} as a local imageblurring operator, although it is often ignored.

Several methods have been proposed to calculate H or its components. These methods can be grouped in three classes: analytical methods, methods based on Monte Carlo simulations, and methods based on measurements. When the system matrix is factored as in Eq. 12.21, various techniques can be combined. To calculate H_{geom}, given the high number of LORs and the complexity of an accurate analytical description, simple models are often used. A simplistic approximation [19] computes the element $(H_{\text{geom}})_{ij}$ as the intersection of the LOR i with voxel j, as depicted in Fig. 12.3 (dashed line). More elaborate models take into account the volume of response (VOR) defined between the faces of the two involved detectors (shadowed area i' in Fig. 12.3) and assign the intersection volume between the voxel and the VOR; to the corresponding matrix element solid angle effects could be included to improve this model. In any case, these models neglect crystal penetration effects: To correct for parallax errors related to the length of the detectors, a VOR including the whole crystal volume should be taken into account (shadowed area i'' in Fig. 12.3), combined with appropriate modelling of this effect.

Since the accuracy of the system model will determine the accuracy of the reconstructed image, we can resort to calculating H as a whole or at least $H_{det.blur}$ by means of Monte Carlo simulations [20] or measurements [21]. Both approaches are time consuming but offer images with higher quality and better spatial resolution.

12.3.6 Ingredient 4: Objective Function

The objective function Φ, also called the *cost function*, is the key ingredient of any iterative algorithm. Φ depends on the unknown image coefficients f_i and on the measured data g: $\Phi = \Phi(f, g)$. Among all possible reconstructed images, the sought image $\hat{f}$ will be the one that maximizes (or minimizes) Φ. This can be expressed as

$$\hat{f} = \arg \max_{f} \Phi. \qquad (12.22)$$

The main role of the objective function is to give a measure of how well an image agrees with the measured data. Depending on how Φ is defined, the objective function can also provide information about how well the image matches some desired image properties (i.e., some prior conditions on the image model). Φ can include a *penalty term $R(f)$* (also called the *regularization term*) imposing some constraints on the reconstructed image. In general, the objective function Φ can be written as

$$\Phi(f,g) = Q(f,g) + \beta R(f), \qquad (12.23)$$

where β is a regularization parameter that controls the balance between the data-fitting criterion and the image property criterion. In PET and SPECT, we can assume that the unknown object consists of different regions in which the radioactive distribution is uniform. Therefore, a roughness penalty term encouraging local smoothness within regions but preserving the edges of the regions can be designed.

12.3.6.1 Maximum A Posteriori Criterion

In a Bayesian framework, the chosen reconstructed image $\hat{f}$ is the one that maximizes the conditional probability prob(f; g), called the *posterior probability distribution*; in other words, the sought solution is the most likely image given the measured data g. The sought image corresponds to maximizing the posterior probability distribution:

$$\hat{f} = \arg\max_{f} prob(f;g). \tag{12.24}$$

The solution is thus called the *maximum a posteriori* (MAP) estimate.

According to Bayes's rule,

$$prob(f;g) = \frac{prob(g;f)prob(f)}{prob(g)}, \tag{12.25}$$

where the conditional probability prob($g;f$) corresponds to the data likelihood given a certain model; this factor tells how well the data agree with the image. Here, prob(f) is the prior probability (or simply prior). This prior distribution reflects the knowledge or assumptions about the properties of those images accepted as solutions (see Sect. 12.3.3). For example, images should not be too noisy. In PET and SPECT we can also assume that images are locally smooth within specific organs or regions, but abrupt changes (sharp edges) are expected between regions.

Since prob(g) is independent of f, this term is dropped in the maximization process of Φ, which reduces to maximizing the numerator prob($g;f$) prob(f) or, equivalently, its logarithm. The cost function is thus reduced to the MAP objective function:

$$\Phi = In\,prob(g;f) + In\,prob(f). \tag{12.26}$$

12.3.6.2 Maximum Likelihood Criterion

The solution $\hat{f}$ is the one for which the measured data have the greatest likelihood according to the conditional probability distribution prob($g;f$). This probability is thus called the *likelihood function* $L(f)$.

We can define our objective function as $L(f)$; the solution that maximizes $L(f)$ is called the maximum likelihood (ML) solution.

$L(f)$ considers the unknown image f as a deterministic parameter vector. Since maximizing a function is equivalent to maximizing its logarithm, the ML solution $\hat{f}$ is obtained by maximizing the log likelihood:

$$\hat{f} = \arg\max_{f} l(f), \tag{12.27}$$

where $l(f) \equiv \ln L(f)$.

The ML criterion can be considered as a limiting case of the MAP criterion when the prior is assumed to be constant, which implies that a priori no solution is preferred over any other. Other authors considered the MAP objective function as a penalized ML estimator. This can be understood comparing Eqs. 12.26 and 12.23. The prior in Eq.12.26 can be viewed as a penalty term.

For a Poisson model, taking into account that the conditional probability prob ($g;f$) is defined by Eqs. 12.15 and 12.16, the log likelihood takes the form

$$l(f) = \sum_{i=0}^{I-1} \ln \mathrm{prob}(g_i; f)$$

$$= \sum_{i=0}^{I-1} g_i \ln \overline{g}_i - \overline{g}_i - \ln g_i! \qquad (12.28)$$

After dropping constants, the objective function $\varPhi$ to be maximized becomes

$$\varPhi = \sum_{i=0}^{I-1} (g_i \ln \overline{g}_i - \overline{g}_i). \qquad (12.29)$$

12.3.6.3   Least Squares Criterion

As mentioned, inverse problems present two difficulties that can be removed together by demanding that we seek, as a solution of the problem, the image that minimizes the difference between the observed data g and the forward-projected image Hf (i.e. the $\hat{f}$ that minimizes $\|g - Hf\|^2$) and that has the smallest norm among all the possible minimizers. This difference is expressed in terms of the Euclidean norm $\|Hf - g\|$, so that the image estimate is

$$\hat{f} = \arg\min_{f} \|Hf - g\|^2. \qquad (12.30)$$

This $\hat{f}$ corresponds to the least squares (LS) solution. In principle, Eq. 12.30 can be solved analytically, and the direct LS solution is

$$\hat{f} = \left(H^T H\right)^{-1} H^T g, \qquad (12.31)$$

where H^T corresponds to the transpose of H. The solution to this minimization problem is unique. This procedure yields an exact solution to the inverse problem if there is one, and a satisfactory approximation if there is not. The linear operator defined by this prescription is the so-called *pseudoinverse* or *Moore–Penrose inverse* $H^+ \equiv (H^T H)^{-1} H^T$.

However, the matrix H is often *ill-conditioned*. As mentioned, the noise components present in g are amplified by the inversion, so that the reconstructed image $\hat{f}$ is not useful. In this case the Moore–Penrose inversion (described in Eq. 12.31) yields a result that is mathematically correct but practically unusable. We conclude that in such a situation the exact inversion must be replaced by something else to keep the reconstruction error under control.

To achieve this, the Moore–Penrose inverse is *regularized* by replacing H^+ by a family of matrices T_γ with the property $T_\gamma g \to H^+ g$ for all g as $\gamma \to 0$. Furthermore, we need a function $\gamma(\varepsilon)$ for which $\|T_{\gamma(\varepsilon)}\| \varepsilon \to 0$ as $\varepsilon \to 0$. With these two ingredients, it is easy to show that for projections contaminated by noise, $g = g_0 + \varepsilon \xi$, where ξ is a noise vector, $\|T_{\gamma(\varepsilon)} g - H^+ g_0\| \to 0$ as $\varepsilon \to 0$. This means that the error in the reconstruction goes to 0 when the noise level ε goes to 0. Note that in this discussion, which applies to finite-dimensional matrices, the error in the reconstruction eventually goes to 0 even for the exact inversion. This is not necessarily the case for operators on infinite-dimensional spaces. The point here, however, is that a careful choice of T_γ and $\gamma(\varepsilon)$ can make the error go to 0 much faster than for the exact inversion. This is particularly important since one usually cannot choose the noise level at will but must take whatever is available. Unfortunately, choosing good regularization parameters is not easy and is more of an art than a science.

An example of such a regularization technique is the *Tikhonov regularization*. It consists of replacing the prescription for the Moore–Penrose inverse by finding the minimizer of the functional $\|g - Hf\|^2 + \gamma \|f\|^2$ with $\gamma > 0$. If the error level in g is known, an appropriate value of γ can be estimated; however, this depends on the particular problem at hand. One can show that the operator defined by this minimization is given by $T_\gamma = (H^T H + \gamma \mathbf{1})^{-1} H^T$.

More generally, regularization can be achieved by adding regularization terms $R(f)$ to the objective function $\Phi(f, g)$, which in our case here is $\|g - Hf\|^2$. In this sense the additional terms in the MAP framework can also be seen as regularization methods to alleviate the ill-posed nature of the reconstruction problem.

12.3.7 Ingredient 5: Optimization Algorithm

In the preceding paragraphs, we presented different criteria to select the "best" image. Given a criterion, finding the optimal solution of an numerical algorithm, typically iterative, is required. The algorithm is designed to produce a sequence of estimates $\hat{f}^1, \hat{f}^2, \ldots, \hat{f}^k, \ldots$ that converges to a solution $\hat{f}$, which optimizes the objective function.

The general structure of most iterative reconstruction algorithms is illustrated in Fig. 12.4. The process begins with $\hat{f}^0$, an initial estimate of the unknown image; $\hat{f}^0$ is usually a uniform image in which all voxels contain the same number of counts (uniform distribution). To speed up the process, the iterative process can be started using the results of a previous reconstruction using FBP, for instance.

The next step consists of forward projecting the current estimate into the measurement space. This is carried out using the system matrix H; the vector $\hat{g}^0 \equiv H\hat{f}^0$ represents an estimate of the data that the detection system would have measured if $\hat{f}^0$ would describe the true object. It follows the comparison between the measured data g and the estimated data $\hat{g}^0$. When the two vectors do not coincide, a set of weights is generated in the measurement space, accounting for the difference. These weights are backprojected using the transpose of H to generate a set of weights into the image space. The latter are then used to modify the last image estimate, which becomes the new estimate $\hat{f}^1$. The starting point of the second iteration thus will be $\hat{f}^1$. This process is repeated again and again until convergence is reached or the process is terminated by the user.

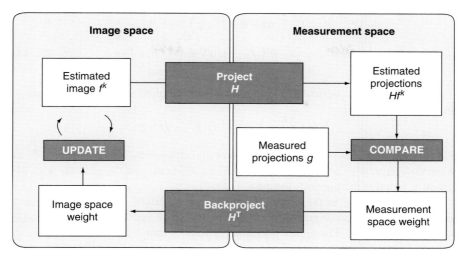

Fig. 12.4 Schematic representation of the iterative process

In principle, the optimization algorithm should asymptotically converge to a global maximum (minimum); it should also be fast stable, and independent of the initial estimate; the final image should not depend on the choice of the numerical algorithm but on the objective function alone. In practice, a large number of iterations is often required to reach convergence; therefore it is common to terminate an iterative algorithm before convergence. As a consequence, different algorithms will lead to different images. In the literature, several optimization methods have been presented. The most popular ones can be classified into search algorithms for functional minimization and functional substitution methods.

12.3.7.1 Search Algorithms

A general expression to describe the search algorithm family of methods is

$$\hat{f}^{(k+1)} = \hat{f}^{(k)} + \alpha_k s^{(k)}, \tag{12.32}$$

which can be interpreted as follows: Choose a search direction $p^{(k)}$ in the image space and take a step α_k in that direction to update the last estimate. Several algorithms can be summarized using this equation. They mainly differ in the procedure for defining the set of directions $\{s^{(k)}\}$ and the rule for determining the step length α_k.

A particular case of this family of methods is the *steepest-ascent algorithm*, which sets $s^{(k)}$ to the gradient of the cost function for the current estimate k, that is, $\nabla \Phi \, (\hat{f}^k, g)$, while α_k maximizes the cost function along the gradient direction. Then, Eq. 12.32 takes the form

$$\hat{f}^{k+1} = \hat{f}^k + \alpha_k \nabla \Phi \left(\hat{f}^k, g \right).$$

$$\alpha_k = \underset{f}{argmax} \, \Phi \left(\hat{f}^{(k)} + \alpha \nabla \Phi \left(\hat{f}^{(k)}, g \right), g \right) \qquad (12.33)$$

This method takes a particularly simple form when applied to find the LS solution. In that case,

$$\nabla \Phi \left(\hat{f}^k, g \right) = H^T H \hat{f}^k - H^T g$$

A faster alternative is the *conjugate gradient algorithm*, which uses $s^{(k)} \equiv p^{(k)}$ as a search vector, with $p^{(k)}$ a vector related to a combination of the gradient with the gradient found at the previous location.

As an alternative to the aforementioned gradient-based methods, *coordinate-ascent algorithms* update each voxel in turn to maximize the objective function with respect to that voxel. The update equation for voxel j then has the form

$$\hat{f}^{k+1}_j = \hat{f}^k_j + h_k. \qquad (12.33)$$

and all other components are left unchanged at this iteration.

12.3.7.2 Functional Substitution Methods and Maximum Likelihood-Expectation Maximization

The functional substitution methods and maximum likelihood-expectation maximization (MLEM) class of methods is based on replacing the original cost function $\Phi(f, g)$ at each step by an alternative or *surrogate* function $\tilde{\Phi}(f, f^k, g)$.

In ET, the best-known algorithm of this kind is the EM algorithm, which offers a general methodology to maximize the likelihood for a probabilistic model. The EM algorithm requires specification of a set of "complete" but unobserved data (in opposition to the measured but "incomplete" data). This unobserved data set can consist of a missing data set or be just an abstraction such that, if these data were accessible, the solution would be easy to find.

In ET, the EM framework is used for finding ML solutions for a Poisson data model; the resulting algorithm is therefore called the MLEM algorithm [22, 23] and is the most widely used algorithm in PET and SPECT.

The complete data set in ET is denoted by the set of random variables $\{w_{ij}; i = 0, \ldots I - 1, j = 0, \ldots, J-1\}$, where I and J are the number of LORs and voxels, respectively. The element w_{ij} represents the number of emissions originating in voxel j and detected in the LOR i, and it is obviously unobserved. (If we had access to w_{ij}, the reconstruction problem would be easily solved).

MLEM consists of two alternating steps, which are repeated until convergence: In the E step, the conditional expectation of the log likelihood of the unobserved data set is calculated, while in the M step this conditional expectation is maximized with respect to the image. Mathematically,

E step:

$$\tilde{\Phi}(f, f^k, g) = E[L(w; f)|g; f^k] \qquad (12.34)$$

M step:

$$f^{(k+1)} = \arg\max_f \tilde{\Phi}\left(f, f^k, g\right) \qquad (12.35)$$

The MLEM is considered a functional substitution method because maximization (the M step) affects the surrogate function $\tilde{\Phi}(f, f^k, g)$, which is previously calculated in the E step.

Taking into account Eqs. 12.15, 12.16, 12.34, and 12.35, the following iterative formula can be derived:

$$\hat{f}_j^{(k+1)} = \frac{\hat{f}_j^{(k)}}{\sum_{i=0}^{I-1} H_{ij}} \sum_{i=0}^{I-1} \frac{g_i}{\sum_{j=0}^{J-1} H_{ij'} \hat{f}_{j'}^{(k)}} H_{ij} \qquad (12.36)$$

This equation can be interpreted as follows: The term $\sum_{j=0}^{J-1} H_{ij'} \hat{f}_{j'}^{(k)}$ corresponds to the forward projection into the measurement space of the last image estimate and can be renamed as $\hat{g}_i^{(k)}$; therefore, the quotient $g_i / \hat{g}_i^{(k)}$ expresses the comparison between the measured data and what we would have measured if the emission source were described by $\hat{f}^{(k)}$ given an imaging system described by H. This ratio, after backprojection into the image space by means of H^T, is used to correct the last image estimate. The normalization term $\sum_{i=0}^{I-1} H_{ij}$ accounts for the detection sensitivity of the system for emissions originating in voxel j.

The MLEM algorithm and its accelerated version, the ordered-subsets expectation maximization algorithm [24] (OSEM), can be considered the most commonly implemented reconstruction methods in ET scanners. The main advantages of MLEM over other iterative algorithms are guaranteed convergence, simple formula, and ease in programming; it does not require any step size or additional parameter, and it uses a realistic statistical model (Poisson) to describe the data distribution. On the other hand, MLEM presents two main shortcomings: The convergence is slow and image noise increases with increasing number of iterations. The latter fact arises from the ML criterion, which seeks the solution having the greatest consistency with the data. But since the data g are noisy in nature, the comparison step $g_i / \hat{g}_i^{(k)}$ encourages those images that better reproduce the noisy measurement.

12.3.8 Correction of Image Degradation Effects

One of the main advantages of iterative reconstruction methods is that the acquisition process, including linear image degradation effects such as attenuation, scatter, collimator or detector response, and so on, can be incorporated within the system matrix. However, accurate modeling of these physical phenomena within the system matrix can become difficult and time consuming. Other effects, such as accidental coincidences in PET, are a nonlinear function of the activity distribution and cannot be included in the model described by Eqs. 12.19 and 12.20. Moreover, although object scatter could be included within the system matrix, this approach would lead to an object-dependent, nonsparse matrix that is difficult to handle and that significantly increases the computational cost. Therefore, object scatter and, in PET, random coincidences can be modeled apart, as additive terms in Eq. 12.20. For PET, this equation becomes

$$\bar{g} = Hf + \bar{s} + \bar{r}, \tag{12.37}$$

where $\bar{s}$ and $\bar{r}$ represent the mean of the scatter coincidences and random coincidences for LOR i, respectively. (In SPECT, the term $\bar{r}$ does not exist.)

Since each of both random and scatter components can be described by a Poisson distribution, an algorithm such as MLEM can be reformulated as

$$\hat{f}_j^{(k+1)} = \frac{\hat{f}_j^{(k)}}{\sum_{i=0}^{I-1} H_{ij}} \sum_{i=0}^{I-1} \frac{g_i}{\hat{g}_i^{(k)} + \hat{s}_i + \hat{r}_i} H_{ij}, \tag{12.38}$$

where $\hat{s}_i$ and $\hat{r}_i$ are the estimates of the scatter and random distributions for LOR i, respectively. (A detailed description of scatter and random estimation techniques is outside the scope of this chapter.) An example of random correction using Eq. 12.37 is shown in Fig. 12.5.

In practice, some scanners, when operating in standard mode, perform online correction of the measured counts by subtracting the estimated randoms and scatter coincidences. This approach not only reduces processing time and data storage but also modifies the statistical distribution of the data. Given a noncorrected measured data vector $\hat{g}$ and a random estimate $\hat{r}$, the precorrected data are now $\hat{g} - \bar{r}$, which is no longer a Poisson variable. The scatter component is usually subtracted as well, so the precorrected measured vector becomes $\hat{g} - \hat{r} - \hat{s}$.

The effect of precorrecting can be described, for example, through a shifted Poisson model [25], a weighted least squares (WLS) estimator [26], or penalized WLS [14].

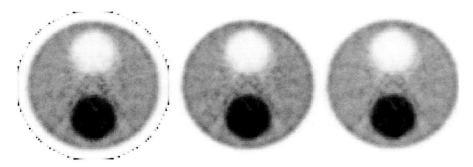

Fig. 12.5 Maximum likelihood-expectation maximization (MLEM) applied to simulated positron emission tomographic (PET) data. *Left*: No correction applied. *Center*: True coincidences only. *Right*: Modified MLEM including random estimate. (Courtesy of I. Torres-Espallardo)

References

1. Radon J.H. Über die Bestimmung von Funktionen durch ihre Integralwerte längs gewisser Mannigfaltigkeiten. Akad. Wiss. 69, 262–277 (1917).
2. Gaskill, J.D. *Linear Systems, Fourier Transforms, and Optics*. Wiley, New York. ISBN 0-471-29288-5 (1978).
3. Hsieh, J. *Computed Tomography: Principles, Design, Artifacts and Recent Advances*. SPIE Press, Bellingham. ISBN 0-8194-4425-1 (2003).
4. Natterer, F. and Wübbeling, F. *Mathematical Methods in Image Reconstruction*. Society for Industrial and Applied Mathematics, Philadelphia, PA. ISBN 0-89871-472-9 (2001).
5. Xu, Y., Tischenko, O. and Hoeschen, C. Approximation and reconstruction from attenuated Radon projections. SIAM J. Numer. Anal. 45 (1), 108–132 (2007).
6. Barret, H.H. and Myers, K.J. *Foundations of Image Science*. Wiley, Hoboken. ISBN 0 471 15300 1 (2004)
7. Bruyant, Ph.P. Analytic and iterative reconstruction algorithms in SPECT. J. Nucl. Med. 43, 1343–1358 (2002).
8. Defrise, M., Kinahan, P.E. and Michel, Ch. J. Image reconstruction algorithms in PET. In: Bailey, D.L., Townsend, D.W., Valk, P.E., and Maisey, M.N., editors. *Positron Emission Tomography: Basic Sciences*. Springer, London. ISBN 1 852337982 (2005).
9. Hutton, B.F., Nuyts, J. and Zaidi, H. Iterative reconstruction methods. In: Zaidi, H., editor. *Quantitative Analysis in Nuclear Medicine*. Springer Science + Business Media, New York. ISBN 0 38 238549 (2006).
10. Leahy, R.M. and Qi, J. Statistical approaches in quantitative positron emission tomography. Statis. Comput. 10, 147–165 (2000)
11. Lewitt, R.M. and Matej, S. Overview of methods for image reconstruction from projections in emission computed tomography. Proc. IEEE 91, 1588–1611 (2003).
12. Qi, J. and Leahy, R.M. Iterative reconstruction techniques in emission computed tomography. Phys. Med. Biol. 51, R541–R578 (2006)
13. Wernick, M.N. and Lalush, D.S. Iterative image reconstruction. In: Wernick, M.N. and Aarsvold, J.N., editors. *Emission Tomography: The Fundamentals of PET and SPECT*. Elsevier Academic Press, London. ISBN 0 12 744482 3 (2004).
14. Fessler, J. Penalized weighted least-squares image-reconstruction for positron emission tomography. IEEE Trans. Med. Imaging. 13, 209–300 (1994).
15. Herman, G.T. *Image Reconstruction from Projections: The Fundamentals of Computerized Tomography*. Springer, New York. ISBN 978-1-85233-617-2 (2009).
16. Matej, S. and Lewitt, R.M. Practical considerations for 3-D image reconstruction using spherically symmetric volume elements. IEEE Trans. Med. Imaging. 15, 68–78 (1996).
17. Buonocore, M.H., Brody, W.R. and Macovski, A. A natural pixel decomposition for two-dimensional image reconstruction. IEEE Trans. Biomed. Eng. BME–28, 69–78 (1981).
18. Herbert, T, Leahy, R. and Singh, M. Fast MLE for SPECT using an intermediate polar representation and a stopping criterion. IEEE Trans. Nucl. Sci. 35, 615–619 (1988).
19. Siddon, R.L. Fast calculation of the exact radiological path for a three-dimensional CT array. Med. Phys. 12, 252–255 (1985).
20. Rafecas, M., Mosler, B., Dietz, M., Pögl, M., Stamatakis, S., McElroy, D.P. and Ziegler, S.I. Use of a Monte–Carlo based probability matrix for 3D reconstruction of MADPET–II data. IEEE Trans. Nucl. Sci. 51, 2597–2605 (2004).
21. Panin, V.Y., Kehren, F., Michel, C., and Casey, M. Fully 3-D PET reconstruction with system matrix derived from point source measurements. IEEE Trans. Med. Imaging. 25, 907–921 (2006).
22. Lange, K. and Carson, R. EM reconstruction algorithms for emission and transmission tomography. J. Comput. Assist. Tomogr. 8, 306–316 (1984).
23. Shepp, L.A. and Vardi, Y. Maximum likelihood reconstruction for emission tomography. IEEE Trans. Med. Imaging. 1, 113–122 (1982).

24. Hudson, H.M. and Larkin, R.S. Accelerated image reconstruction using ordered subsets of projection data. IEEE Trans. Med. Imag. 13, 601–609 (1994).
25. Yavuz, M. and Fessler, J.A. Statistical image reconstruction methods for randoms-precorrected PET scans. Med. Image Anal. 2, 369–378 (1998).
26. Huesman, R.H., Gullberg, G.T., Greenberg, W.L., and Budinger, T.F. Users manual – Donner algorithms for reconstruction tomography. Publication PUB-214, Lawrence Berkeley Laboratory, Berkeley, (1977).

Chapter 13
Biokinetic Models for Radiopharmaceuticals

Augusto Giussani and Helena Uusijärvi

13.1 Internal Dosimetry of Radiopharmaceuticals

Radiopharmaceuticals administered for diagnostic or therapeutic applications in humans are selectively transported into specific organs and tissues of the body, metabolised, and finally excreted according to their biochemical and metabolic properties. Due to the presence of the radioactive label, each of the body regions containing the substance becomes an emitting source (*source region*), which can also irradiate the neighbouring tissues (defined as *target regions*). Consequently, each body organ or tissue could receive a radiation dose (*absorbed dose*) after administration of radiopharmaceuticals even if no activity is present in it. The absorbed dose delivered by incorporated radioactive material is called the *internal dose*. Direct measurements of the internal dose are not possible for evident practical reasons, so this quantity has to be calculated using a mathematical approach. Such an approach has to take into account that

- The activity in a source region varies with time not only because of the physical process of nuclear transformation but also due to the metabolic behaviour of the carrier molecule.
- The energy of the radiation emitted by a radionuclide present in a given source region may be fully or partially absorbed in the source region itself or partially deposited in the neighbouring tissues (depending on type and energy of the emitted radiation as well as the relative positions of the pair source–target regions).
- The absorbed dose to a given target region is the sum of the absorbed doses delivered separately by each source region.

A. Giussani (✉)
Helmholtz Zentrum München, Department of Medical Radiation Physics and Diagnostics, 85764 Neuherberg, Germany
e-mail: augusto.giussani@helmholtz-muenchen.de, agiussani@bfs.de

H. Uusijärvi
Medical Radiation Physics Lund University, Malmö, Sweden

M.C. Cantone and C. Hoeschen (eds.), *Radiation Physics for Nuclear Medicine*,
DOI 10.1007/978-3-642-11327-7_13, © Springer-Verlag Berlin Heidelberg 2011

Such an approach was formalised in 1988 by the MIRD (Medical Internal Radiation Dose) Committee of the U.S. Society of Nuclear Medicine with the publication of the Primer for internal dose calculations [1]. According to that publication, the absorbed dose D_{r_k} in a target region r_k is given by

$$D_{r_k} = \sum_h \tilde{A}_h \, S(r_k \leftarrow r_h) \tag{13.1}$$

where $\tilde{A}_h$ (called the *cumulated activity*) is the number of nuclear transformations that occurred in the source region r_h (time integral of the activity curve) and

$$S(r_k \leftarrow r_h) = \frac{k \sum_i n_i E_i \phi_i (r_k \leftarrow r_h)}{m_{r_k}} \tag{13.2}$$

is the so-called S factor (the same as the dose factor [DF] used by other authors), which takes into account the physical properties of the radionuclide (n_i is the emission probability of the ith radiation, E_i is the energy of the ith radiation) and of the anatomy of the subject (m_{r_k} is the mass of the target region r_k, $\phi_i (r_k \leftarrow r_h)$ is the fraction of the energy of the ith radiation emitted in r_h that is absorbed in r_k). Here, k is a multiplicative factor that depends on the units used.

For practical reasons, the absorbed dose per unit administered activity is often used; in that case, Eq. 13.1 is rewritten as

$$\frac{D_{r_k}}{A_0} = \sum_h \tau_h \, S(r_k \leftarrow r_h) \tag{13.3}$$

where A_0 is the administered activity and $\tau_h = \frac{\tilde{A}_h}{A_0}$ is the number of transformations per unit activity. In this expression, τ_h has been called the *residence time* since it has the units of time and indeed corresponds to the residence time in the case of monoexponential clearance from the organ where activity was administered. However, in more general cases this terminology can be somewhat deceptive, and the ratio between the cumulated activity (integral of the activity curve) and the initial activity should not be confused with the biokinetic characteristic times described later in this chapter.

The International Commission on Radiological Protection (ICRP) has adopted a similar approach in its publications "Radiation Dose to Patients from Radiopharmaceuticals" [2–4] but using slightly different terminology:

$$D_T = \sum_S \tilde{A}_S \, S(T \leftarrow S) \tag{13.4}$$

with

$$S(T \leftarrow S) = \frac{c \sum_i Y_i E_i \varphi_i (T \leftarrow S)}{M_T} \tag{13.5}$$

having substituted S for r_h, T for r_k, Y_i for n_i, and φ_i for ϕ_i. The use of alternative symbols (which also differ from the ones used by ICRP for occupational exposures) can be confusing, and recently a proposal has been made for standardisation of nomenclature in radiopharmaceutical dosimetry. In MIRD Pamphlet 21 [5], it is suggested that source and target regions are indicated by the symbols r_S and r_T, respectively; the number of transformations in a source region is called *time-integrated activity* and indicated with the symbol $\tilde{A}(r_S, T_D)$ (where T_D is the time of integration); and the number of transformations per unit intake is called *time-integrated activity coefficient* and is given the symbol $\tilde{a}(r_S, T_D)$. This new nomenclature is adopted in the rest of this chapter.

Irrespective of the terminology used, however, the process of estimation of the internal dose can be basically described as a combination of three inputs: one physical input (probabilities and energies of the emitted radiations), one "physicoanatomical" input (masses of the target regions and absorbed energy fractions), and one biokinetic input (the number of transformations in the source region).

The physical properties of the radionuclides are in general well known from nuclear physics studies, and a compilation of the relevant information for internal dosimetry has been issued by ICRP [6]. The calculation of the fraction of deposited energy for each pair of source–target regions is dealt with in Chap. 14 of this book. The present chapter describes the modalities for the quantitative determination of the activity in the organs of the patients (Sect. 13.2); discusses the modelling approaches used for the determination of the time–activity curves $A(r_S, t)$, which are the starting point for the calculation of the time-integrated activity $\tilde{A}(r_S, T_D)$ (Sect. 13.3); and presents some practical applications (Sect. 13.4).

13.2 Measurement of Activity Curves in Patients

The uptake, distribution, and retention of the administered activity in the organs and tissues of patients can be studied by acquiring gamma camera, single-photon emission computed tomographic (SPECT), or positron emission tomographic (PET) images over the body region of interest (ROI) at different time points after the activity injection. It is common to collect blood samples as well as urine and sometimes faeces samples depending on the excretion pathway of the radiopharmaceutical.

13.2.1 Data Collection

When using a radiopharmaceutical suitable for gamma cameras, such as radiopharmaceuticals containing ^{99m}Tc, ^{111}In, or ^{123}I, planar gamma camera or SPECT images can be collected. Acquisition times depend on the injected activity: With lower injected activity, the measured count rate will be lower, and a longer time is needed to get satisfying counting statistics in the images. One planar whole-body image is built from several fields-of-view (FOVs), each called a *bed position*. One image acquisition takes around 15–20 min, depending on the time of each bed position. One rotation around the patient when collecting SPECT images may take up to about 40 min, depending on the number of projections and duration of each projection. Due to the long imaging time for SPECT, it is not possible to make whole-body SPECT images. In biokinetic studies for determination of the time–activity curves in the source regions, the most common procedure is thus to collect planar whole-body images at different time points after the injection. This procedure can be combined with one SPECT image taken over the body ROI (i.e., over the organs of interest for the absorbed dose calculations). The SPECT image is used for the absolute activity quantification, and the series of planar images is used for the shape of the time–activity curves. The requirement for sufficiently elevated count rates is particularly important in the case of SPECT acquisition since with too few counts the final image will be too noisy, and the activity quantification will be unreliable.

When using a positron-emitting radiopharmaceutical (e.g., ^{18}F or ^{11}C), PET images are suitable for determining the activity in the different organs. In contrast to SPECT cameras, PET cameras contain a ring of detectors and can therefore measure the radiation coming from the patient in all directions at the same time. This shortens the imaging time tremendously with respect to SPECT, and a whole-body PET image takes around 25 min, depending on the time per bed position and the number of bed positions needed.

In planar gamma camera, SPECT, as well as PET images, it is possible to draw ROIs and obtain the number of counts per second (cps) in a specific part of the image. SPECT and PET images are tomographic images in which each organ is represented as a series of consecutive slices of known thickness. To determine the activity content in an organ, one needs to draw one ROI in each of the slices containing the organ. The counts per second in each slice i are added to obtain the total number of counts per second in the organ:

$$C_{\text{tot}} = \sum_{i=1}^{k} C_i \tag{13.6}$$

where k is the number of slices.

13.2.2 Tracer and Tracee

The ideal scenario, of course, is to collect biokinetic data using the radionuclide of interest. However, in therapeutic applications of radiopharmaceuticals, this is not always possible. For example, ^{90}Y, a commonly used radionuclide for systemic therapy, does not emit photons suitable for imaging, and a substitute for this radionuclide is needed when performing biokinetic studies. In most cases, the distribution and biological retention of the radiopharmaceutical are determined by the chemical compound and not by the attached radioactive labels, so the biokinetic information can be obtained using labels that are suitable for gamma camera, SPECT, or PET imaging (like ^{111}In or ^{86}Y as a substitute for ^{90}Y). In this case, ^{90}Y is called the *tracee* (i.e., the substance that is to be studied), and ^{111}In and ^{86}Y are called *tracers* (i.e., the substances used to trace the biokinetics of the tracee). The same is valid for many other alpha- and beta-emitters used in clinical radionuclide therapy.

13.2.3   Efficiency Calibration

To convert the number of counts per second in an ROI to the activity content in the organ, it is necessary to make an efficiency calibration of the camera. The efficiency is usually given as counts per second per becquerel, that is, the number of counts that the camera registers per decay. The efficiency calibration can be made by measuring a known activity, drawing ROIs around the source on every slice of the image, and summing the counts per second, correcting them for physical decay, and dividing by the known activity:

$$\varepsilon = \frac{C_{tot}^*}{A_0} \tag{13.7}$$

where ε is the efficiency of the camera, C_{tot}^* is the number of total counts per second corrected for physical decay, and A_0 is the activity in the source region at the initial time point. It is important to perform the efficiency calibration for different activities in the source to make sure the response of the camera is linear. For planar gamma camera images, the efficiency is determined in the same way by drawing an ROI around the source on the planar image and correcting the counts for physical decay. However, planar whole-body images make it possible to determine the percentage of injected activity directly without using the efficiency ε of the camera. The counts per second in the whole body at the first acquisition can be seen as 100% of the injected activity. It is, however, important to make sure that the patient has not voided the bladder before the acquisition. This method can only be used if all activity injected is left within the body. By dividing the counts per second in the organs (corrected for background and attenuation; see Sects. 13.2.4 and 13.2.5 for more details) with the counts per second in the whole body from the first acquisition, the percentage injected activity (%IA) is obtained:

$$\%IA_{organ} = \frac{C_{organ}}{C_{wholebody}} \tag{13.8}$$

This expression gives the percentage activity in the organ of interest without knowing the efficiency factor and the injected activity. For dose estimates, however, this information is crucial. Before injection, the activity is therefore measured in an activity meter, usually a well ionisation chamber. It is therefore important not only that the gamma, SPECT, and PET cameras are calibrated with respect to efficiency, but also that the activity meter is calibrated carefully. When using radionuclides emitting photons with high energy (e.g., PET substances with annihilation photons of 511 keV), the measurement geometry is not crucial due to the low self-absorption in the source. However, when using radionuclides with lower photon energy (e.g., ^{99m}Tc with 140 keV and ^{123}I with 159 keV), the self-absorption in a source (e.g., a syringe or vial) will be dependent on the shape and volume of the source; it is therefore important to calibrate the activity meter for the specific source geometry that will be used. The same is true for the gamma counter in which blood and urine samples can be measured. It is also important to make an efficiency calibration using the geometry that is intended to be used for the gamma counter.

13.2.4 Background Correction

The SPECT and PET images consist of thin slices with no overlapping tissue, and there will be no need for background correction. On the other hand, in the planar gamma camera images, the whole body will be shown in a two-dimensional image, and there will be overlapping tissue both in front and behind many of the organs. It is therefore necessary to make background corrections. A ROI should be placed at a location of the body where the activity uptake is expected to be similar to the activity in the overlapping tissue (e.g., in the leg of the patient where there is no specific activity uptake or close to the organ of interest). The counts per second in the background ROI are subtracted from the counts per second in the ROI of the organ to obtain the net counts per second in the ROI of the organ:

$$C_{net} = C_{organ\ ROI} - C_{bkg\ ROI}.$$ (13.9)

13.2.5 Attenuation Correction

For SPECT and PET systems, the number of counts per second received is usually automatically corrected for attenuation using a computed tomographic (CT) scanner built in the SPECT or PET unit or using a transmission source belonging to the unit. Some PET systems give the activity concentration in an ROI, but in other systems as well as in SPECT images the user needs to convert the counts per second to the activity by hand. The planar gamma camera images are usually not corrected for attenuation, and the counts per second registered by the camera will be dependent on the depth of the body at which the organ is located. The attenuation-corrected counts per second in an organ can be expressed as

$$C_{attcorr} = C_{net}e^{\mu_{eff}d}$$ (13.10)

where μ_{eff} is the effective attenuation coefficient, and d is the depth of the organ. The thickness of the organ is assumed to be infinitesimal. Values of the effective attenuation coefficient μ_{eff} are tabulated. For more specific applications, values of μ_{eff} can be directly estimated in the real experimental conditions by placing a planar

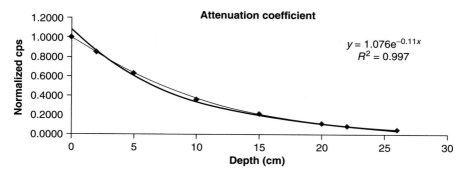

Fig. 13.1 Determination of the effective attenuation coefficient. A 50-mL cylindrical source with a diameter of 3 cm was placed in a 30 × 30 × 30 cm³ water phantom. The effective attenuation coefficient was 0.113 cm⁻². These values correspond to the gamma camera head above the phantom; the effective attenuation coefficient will be slightly higher for the gamma camera head under the table due to attenuation in the table itself. *cps*: counts per second

source in a water phantom, varying the depth, plotting the counts per second as a function of depth, and fitting an exponential function, as seen in Fig. 13.1.

The depth of the organ is difficult to know if there is no access to a CT image (in some cases, the acquisition of a magnetic resonance image may be possible), but there is a way to get around this: the conjugate view method [7]. In the conjugate view method, two 180° opposed images are used, usually one anterior (from the front) and one posterior (from the back). The counts per second from the anterior view can be expressed as

$$C_A = \varepsilon A e^{-\mu_{\text{eff}}d} \tag{13.11}$$

where A is the activity in the organ. The counts per second in the posterior view can be expressed as

$$C_P = \varepsilon A e^{-\mu_{\text{eff}}(T-d)} \tag{13.12}$$

where T is the thickness of the patient at the particular part of the body. In the conjugate view method, the geometric mean of the anterior and posterior counts per second is used:

$$\sqrt{C_A C_P} = \sqrt{\varepsilon^2 A^2 e^{-\mu_{\text{eff}}(d+T-d)}} = \varepsilon A e^{-\mu_{\text{eff}}\frac{T}{2}} \tag{13.13}$$

The organ depth is cancelled out, and only the knowledge of the total thickness of the patient is needed. This equation can be rewritten so that the activity A in the organ is expressed as

$$A = \frac{\sqrt{C_A C_P}}{\varepsilon} e^{\frac{\mu_{\text{eff}}T}{2}} \tag{13.14}$$

13.2.6 Time Schedule

The time after the activity injection when the images, blood, urine, and faeces samples should be collected depends on the radiopharmaceutical studied. For example, peptides are known to have a short retention in the blood, while antibodies usually circulate with the blood for a longer time. According to the data shown in Fig. 13.2, it is important to take early blood samples when studying a peptide to correctly evaluate the fast clearance from the circulation: If the first blood sample would be taken 24 h after injection, it would be impossible to evaluate how fast the peptide has been eliminated from the circulation. On the other hand, for antibodies it probably would not make a big difference if the early blood samples were left out and the first sample was taken 24 h after injection. In this case, it is important to include late blood samples to obtain a correct estimation of the activity remaining in the blood: If the sample collection would be stopped 24 h after injection, it would not be possible to evaluate the long-term retention in the blood.

The time points for imaging depend on the effective half-life of the radiopharmaceutical studied, that is, a combination of the physical half-life and the biological half-life. The study of previously published biokinetic data of similar radiopharmaceuticals in humans or animals can work as a guide when deciding the imaging time scheme [7].

A general protocol could include one acquisition around the time of the effective half-life and two additional acquisitions after around three and five effective half-lives. To make it possible to solve the compartment models (see the next section), at least three data points are needed for each exponential term of the model [7]. For rapid clearance of the radiopharmaceutical, it is important to include early time points so the area under the curve is not overestimated. To characterise the long-term retention and not underestimate the area under the curve, late time points are necessary.

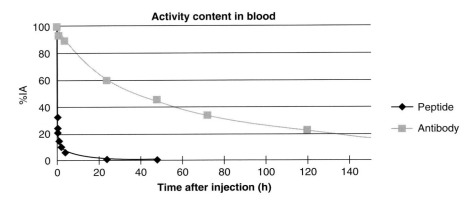

Fig. 13.2 The activity content in blood at different times after injection. The peptide refers to ^{177}Lu-DOTATOC and the antibody to ^{177}Lu-Rituximab. *%IA* percentage injected activity (Data from Dr. F. Forrer at Basel University Hospital, Switzerland)

It might be necessary to collect urine and faeces samples, depending on the excretion route of the radiopharmaceutical. It is more common to collect urine samples than faeces samples, not only because the kidneys are a common path of excretion but also because of the inconvenience for patients to collect faeces. When collecting urine samples, the cumulated activity is of interest. The patients can be asked to collect each bladder voiding in a separate sample, divide the urine samples into different time intervals, or collect all urine in one sample. The total duration of urine collection depends on the effective half-life of the radiopharmaceutical in the body.

13.3 Formulation and Identification of Biokinetic Models

13.3.1 Analytical Description of the Activity Measurements

The activity values measured as described in the previous section are the basis for the definition of the time–activity curves in the source organs. The simplest way to obtain an analytical formula for the time-dependent activity $A(r_S, t)$ in the source region r_S is to fit a sum of exponentials to the experimental data:

$$A(r_S, t) = \sum_{i=1}^{n} a_i e^{-b_i t}. \tag{13.15}$$

In general, a sum of exponentials is able to describe the uptake–clearance curves measured in patients' organs. The only condition is to have a number of experimental data that exceeds the number of free coefficients to be fitted. Commonly, the initial conditions at time $t = 0$ are known; that is, there is no initial activity in the organ r_S unless it is the organ where the activity A_0 is administered. So, the following constraint equations need to be considered:

$$A(r_S, 0) = \sum_{i=1}^{n} a_i = 0 \tag{13.16a}$$

or, alternatively,

$$A(r_S, 0) = \sum_{i=1}^{n} a_i = A_0 \tag{13.16b}$$

and the number of free parameters becomes $2n - 1$.

Such an approach is easy and immediate to implement, and the time-dependent activity curve in the region can be easily integrated over the time T_D to obtain the value of the time-integrated activity:

$$\tilde{A}(r_S, T_D) = \sum_{i=1}^{n} \frac{a_i}{b_i} \cdot (1 - e^{-b_i T_D}) \tag{13.17}$$

242 A. Giussani and H. Uusijärvi

This analysis, however, is not based on physiological considerations and does not take interconnections between organs into account. Consequently, the obtained curves cannot be used in a prospective way (e.g., to predict the future evolution of organ activities or to simulate the influence of modifications in the administration modalities). The curves have indeed been constructed to describe the available experimental data, and their validity cannot be extended over the period of observation.

13.3.2 Definition of Compartmental Models

A more elegant and powerful (but also more critical and time-consuming) approach is that of using a physiologically realistic model to describe the data. In general, compartmental analysis is applied for the development of biokinetic models. A compartmental model consists of a series of interconnected units (the compartments). The fundamental assumption is that the substance investigated behaves uniformly and homogeneously within one given compartment, so its behaviour in that compartment can be described by means of a variable of state. Figure 13.3 shows a simple four-compartment model, with the units (compartments) represented by the boxes and some arrows expressing the exchange of material between compartments or clearance towards the external environment.

In the case of radiopharmaceuticals administered to patients, the compartments may represent the source organs, and the corresponding variables of state are the

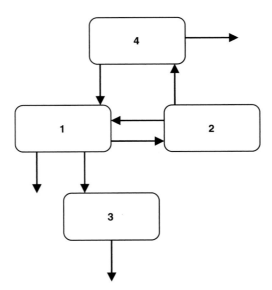

Fig. 13.3 Four-compartment model

activities in those organs. Actually, it is not necessary to have a one-to-one corre-spondence between organs and compartments; one compartment may be taken to represent several organs, where the substance is assumed to behave similarly, or one organ can be split into more compartments when the substance is known not to behave homogeneously within that specific organ.

The advantages of a compartmental approach are manifold: All collected data for all organs and ROIs can be simultaneously analysed with a single structure supposedly developed starting from physiological considerations and taking into account the interdependencies and connections between the different units. So, more general information is used for the formulation of the model in addition to the spe-cific biokinetic data, and its validity may be extended beyond the duration of the experimental study, provided that some theoretical conditions (see Sect. 13.3.4) are verified. Under these assumptions, the model can well be applied for prospective and not only for descriptive purposes.

13.3.3 Mathematical Solution of the Compartmental Models

The mathematical description of compartment models is given by a system of dif-ferential equations, where the variation of the variable of state (e.g., the activity) in compartment i is given by the net difference of the material exchange rates entering that compartment and those leaving it:

$$\frac{dq_i(t)}{dt} = R_{\text{in}}(t, \underline{q}) - R_{\text{out}}(t, \underline{q}) \tag{13.18}$$

having indicated with q_i the generic variable of state of compartment i, with $\underline{q}$ the vector of all variables of state and with R the generic exchange rate of material between compartments. $R(t, \underline{q})$ can depend on time and on the variables of states (and also on some other external parameters, not indicated here). For the specific case of the model in Fig. 13.3,

$$\begin{cases} \dfrac{dq_1(t)}{dt} = -R_{21}(t, \underline{q}) - R_{31}(t, \underline{q}) - R_{01}(t, \underline{q}) + R_{12}(t, \underline{q}) + R_{14}(t, \underline{q}) \\[2mm] \dfrac{dq_2(t)}{dt} = R_{21}(t, \underline{q}) - R_{12}(t, \underline{q}) - R_{42}(t, \underline{q}) \\[2mm] \dfrac{dq_3(t)}{dt} = R_{31}(t, \underline{q}) - R_{03}(t, \underline{q}) \\[2mm] \dfrac{dq_4(t)}{dt} = R_{42}(t, \underline{q}) - R_{14}(t, \underline{q}) - R_{04}(t, \underline{q}) \end{cases}$$

$$\tag{13.19}$$

Here, the engineering notation is used, with R_{ij} meaning exchange from compart-ment j to compartment i, and 0 indicating the external environment.

Several expressions are available to describe the exchange of material between compartments. The simple assumption of first-order kinetics is commonly used in the absence of specific evidence of more complex mechanisms regulating the transfer processes; according to this assumption, the material transferred between compartments (or to the external environment) is proportional to the amount of material present in the parent compartment:

$$R_{ij}(t, \underline{q}) = k_{ij}q_j(t), \tag{13.20}$$

where k_{ij} is a constant transfer rate coefficient. The clearance from compartment i is given by the sum of all outflows from i, and its characteristic time is given by the inverse of the sum of the corresponding transfer rate coefficients:

$$\tau_i = \frac{1}{\displaystyle\sum_{j=0, j\neq i}^{n} k_{ji}} \tag{13.21}$$

(not to be confused with the "deceptive" residence time defined in the original MIRD primer).

The system of differential equations (Eq. 13.19) then becomes

$$\begin{cases} \dfrac{dq_1(t)}{dt} = -(k_{21} + k_{31} + k_{01}) \cdot q_1(t) + k_{12} \cdot q_2(t) + k_{14} \cdot q_4(t) \\[2mm] \dfrac{dq_2}{dt} = k_{21} \cdot q_1(t) - (k_{12} + k_{42}) \cdot q_2(t) \\[2mm] \dfrac{dq_3}{dt} = k_{31} \cdot q_1(t) - k_{03} \cdot q_3(t) \\[2mm] \dfrac{dq_4}{dt} = k_{42} \cdot q_2(t) - (k_{04} + k_{14}) \cdot q_4(t) \end{cases} \tag{13.22}$$

and can be more easily written in matrix form:

$$\frac{d\underline{q}}{dt} = A \cdot \underline{q} \tag{13.23}$$

where A is the compartmental matrix of the transfer rate coefficients:

$$A = \begin{bmatrix} -(k_{01} + k_{21} + k_{31}) & k_{12} & 0 & k_{14} \\ k_{21} & -(k_{12} + k_{42}) & 0 & 0 \\ k_{31} & 0 & -k_{03} & 0 \\ 0 & k_{42} & 0 & -(k_{04} + k_{14}) \end{bmatrix} \tag{13.24}$$

From linear algebra, we know that the solution of Eq. 13.23 is a sum of exponentials:

$$q_i(t) = \sum_{m=1}^{n} a_{mi} e^{-b_m t} \tag{13.25}$$

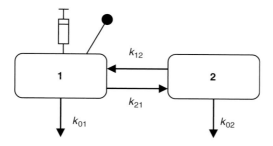

Fig. 13.4 Simple two-compartment model with both the external input (*represented by the syringe*) and the sampling point (*represented by the dot*) in compartment 1

with n the number of compartments in the model. So, we have obtained that the analytical solution of a model with n compartments is a sum of exponentials for which the terms are less than or equal to the number of compartments (depending on the initial conditions, some of the coefficients a_{mi} might be equal to 0).

The relationship between the coefficients a_{mi} and b_m of Eq. 13.25 and the transfer rate coefficients k_{ij} of the compartmental structure, however, is not straightforward. For the sake of simplicity, let us check it for the simple two-compartment model presented in Fig. 13.4.

The syringe represents an external input into compartment 1 (e.g., the administration of a bolus activity A_0 of a radiopharmaceutical at time $t = 0$), and the dot indicates a sampling point (i.e., measurements are performed in compartment 1 at given times postadministration). The solution of Eq. 13.23 for the two-compartment model in Fig. 13.4 is the sum of two exponentials:

$$A(r_{S_1}, t) = q_1(t) = a_{11} \cdot e^{-b_1 \cdot t} + a_{21} \cdot e^{-b_2 \cdot t} \tag{13.26}$$

where

$$a_{11} = A_0 \frac{k_{12} + k_{02} - k_{21} - k_{01} + \sqrt{(k_{12} + k_{02} + k_{21} + k_{01})^2 - 4 \cdot (k_{21}k_{02} + k_{01}k_{12} + k_{01}k_{02})}}{2 \cdot \sqrt{(k_{12} + k_{02} + k_{21} + k_{01})^2 - 4 \cdot (k_{21}k_{02} + k_{01}k_{12} + k_{01}k_{02})}}$$

$$a_{21} = A_0 \frac{k_{21} + k_{01} - k_{12} - k_{02} + \sqrt{(k_{12} + k_{02} + k_{21} + k_{01})^2 - 4 \cdot (k_{21}k_{02} + k_{01}k_{12} + k_{01}k_{02})}}{2 \cdot \sqrt{(k_{12} + k_{02} + k_{21} + k_{01})^2 - 4 \cdot (k_{21}k_{02} + k_{01}k_{12} + k_{01}k_{02})}}$$

$$b_1 = \frac{(k_{12} + k_{02} + k_{21} + k_{01}) - \sqrt{(k_{12} + k_{02} + k_{21} + k_{01})^2 - 4 \cdot (k_{21}k_{02} + k_{01}k_{12} + k_{01}k_{02})}}{2}$$

$$b_2 = \frac{(k_{12} + k_{02} + k_{21} + k_{01}) + \sqrt{(k_{12} + k_{02} + k_{21} + k_{01})^2 - 4 \cdot (k_{21}k_{02} + k_{01}k_{12} + k_{01}k_{02})}}{2}$$

$$\tag{13.27}$$

It is clearly evident that as the number of compartments increases, the complexity of the relation increases. Many current software packages are available that enable solutions of compartmental models in order to obtain the unknown values of the model parameters without the need to solve equations such as 13.27. However, the facility to use these software packages and obtain solutions even for complex models should not be misleading and should not prevent checking the reliability and the theoretical validity of the model. The following discussion gives some brief hints of the important aspects of a priori and a posteriori identifiability.

13.3.4 A Priori Model Identification

To explain the problem of a priori identification, let us consider again the example shown in Fig. 13.4. The model has four unknown parameters: k_{12}, k_{02}, k_{21}, and k_{01}, and there are four equations relating the k_{ij}'s to the fit coefficients a_{11}, b_1, a_{21}, and b_2 (Eq. 13.27). However, the two coefficients a_{11} and a_{21} are linked by the constraint equation $a_{11} + a_{21} = A_0$ (initial activity injected into compartment 1), so that the actually independent equations are three instead of four, and the system in Eq. 13.27 cannot be solved. In other words, there are infinite combinations of the values of k_{ij}'s that correspond to a given set of fit coefficients, and the model is unidentifiable. Figure 13.5 clearly shows the implications of an unidentifiable model. The two-compartment model of Fig. 13.4 was fitted to some experimental data measured in compartment 1. As discussed, there are infinite combinations of the model parameters that correspond to the same best-fit curve. The left panel of Fig. 13.5 shows the curves for two of the infinite sets of parameters; as expected, the curves in compartment 1 are identical. However, the predictions in compartment 2 are significantly different for the two sets of parameters (right panel of Fig. 13.5), and actually any other combination would give a different prediction for the curve in compartment 2. However, the advantage of a model lies in the possibility to use it

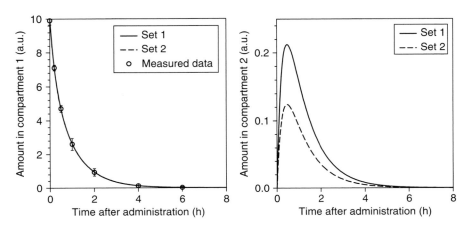

Fig. 13.5 *Left panel*: Fit of the compartmental model of Fig. 13.4 to experimental data measured in compartment 1. Two curves are shown (*curves are not distinguishable*). *Right panel*: Corresponding predictions in compartment 2 for the two sets of parameter solutions

for predicting the evolution of the system without the need to make additional investigations (e.g., at times when or in compartments where experimental measurements are impossible or difficult to perform). An unidentifiable model is useless from this point of view.

The identifiability issue is purely a mathematical issue: No matter how good or how frequent the measurements are, no matter how good the fit to the data is (left panel of Fig. 13.5), it will always be mathematically impossible to obtain unique solutions for the model parameters. For the example shown, there will always be three independent equations for four unknowns, and making more measurements in compartment 1 or improving their accuracy and precision would not help. To be able to obtain the parameter values, more independent equations are needed. This could be done, for example, considering a second sampling point in compartment 2. The activity in this compartment would be given by

$$A(r_{S_2}, t) = q_2(t) = a_{12} \cdot e^{b_1 \cdot t} + a_{22} \cdot e^{b_2 \cdot t} \tag{13.28}$$

so that we now have two further equations (the ones relating the fit coefficients a_{12} and a_{22} to the k_{ij}'s). Of these equations, only one is independent since we also have the further constraint $a_{12} + a_{22} = 0$ (no initial activity in compartment 2); however, one additional independent equation is exactly what we needed to be able to solve the system of equations and find unique solutions for the k_{ij}'s.

This simple example shows that the ability to find unique solutions for the unknown parameters is not a property of the model structure alone but rather of the combination between model structure and proposed experiment (choice of input and output compartments). The identifiability issue can be formalized following the clear and circumstantial approach used by Carson et al. [8].

Let the combination of model structure and proposed experiment be defined as the *constrained structure*:

1. One model parameter of a constrained structure is said to be **unidentifiable** in the observation interval $[t_0, T]$ if there exists an **infinite** number of solutions for that parameter. If **one or more parameters** of a model are **unidentifiable**, then the **model** is said to be **unidentifiable**.
2. One model parameter of a constrained structure is said to be **identifiable** in the observation interval $[t_0, T]$ if there exists a **finite** distinct number of solutions for that parameter. If **all parameters** are **identifiable**, then the **model** is said to be **identifiable**.
3. One model parameter of a constrained structure is said to be **uniquely identifiable** in the observation interval $[t_0, T]$ if there exists **one and only one** solution for that parameter. If **all parameters** are **uniquely identifiable**, then the **model** is said to be **uniquely identifiable**. If at least **one parameter** is **nonuniquely identifiable**, then the **model** is said to be **nonuniquely identifiable**.

The a priori (theoretical) identifiability should always be tested before starting with the experimental studies (data collection). If the structure proves to be unidentifiable, the experiment should be revised. In the previous example, adding a new

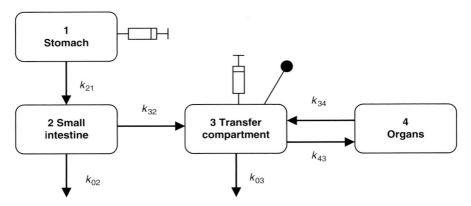

Fig. 13.6 Four-compartment model to study intestinal absorption with the help of the double-tracer technique

sampling site in the second compartment enabled the unique identifiability of all parameters. A second possibility is to change the model structure, making it simpler; for example, eliminating one transfer path in the structure shown in Fig. 13.4 reduces the number of unknown parameters to three, that is, the number of free equations. This simplification of the model is recommended only when it is physiologically realistic and does not implicate a loss of the information that can be obtained with the model. A third option consists in looking for additional information (e.g., in the scientific literature) to be used for the identification of one or more parameters. Let us show that using the constrained model structure shown in Fig. 13.6.

The model corresponds to the description of absorption of ingested material into blood on the basis of the ICRP30 gastrointestinal (GI) tract model [9] (the model used before the introduction of the human alimentary tract model in Publication 100 [10]). Ingested material is transported from the stomach to the small intestine, and from there it can be absorbed into the systemic circulation (path k_{32}) or be directly excreted (path k_{02}). The proposed experiment is a typical double-tracer investigation, with simultaneous administration of one oral and one intravenous tracer and measurement of both tracers in plasma (transfer compartment). It can be shown that the model is identifiable; however, the parameters k_{21}, k_{02}, and k_{32} all have two possible solutions, with k_{21} and k_{02} having exchangeable values. The ambiguity can be solved knowing (from the literature) that gastric emptying is faster than the removal of material from the intestine; so, from the two possible combinations of parameter values, the one for which $k_{21} < k_{02}$ should be chosen. It might be interesting to note that the absorbed fraction, that is, the fraction of material entering the systemic circulation and defined in ICRP30 as the ratio

$$f_1 = \frac{k_{32}}{k_{32} + k_{02}}, \tag{13.29}$$

is a uniquely identifiable combination of nonuniquely identifiable parameters. Parameter lumping is indeed another solution to the identifiability issue.

Checking the identifiability of the chosen constrained model structure is therefore an essential step in the modelling process. However, this is not an easy task, especially for models consisting of more than three compartments. Carson et al. [8] presented some methods of testing for identifiability that are of difficult practical implementation. One necessary condition for identifiability is that each compartment should be linked with at least one external input and at least one sampling point. Although this condition does not guarantee that the structure is identifiable, it enables recognition of at least the unidentifiable ones.

A significant help in checking the identifiability issue could be obtained with specific software; at the moment, only a few such software packages are available [11, 12].

13.3.5   A Posteriori Model Identification

The check of the a priori identifiability is only the first step in solving the model; once the uniqueness of the solutions for the model parameters has been verified from a theoretical (mathematical) point of view, the parameter values have to be determined. This is usually done by fitting the model curves to the experimental data by maximum likelihood or least squares minimization techniques. Different from the issue of a priori identifiability, the fitting techniques are described in detail in many texts and Web resources. Here, we concentrate only on a few aspects related to the quality of the results obtained.

First, it should be stressed again that the a priori identifiability (necessary condition for having unique values of the model parameters) does not guarantee that the model is the correct one to describe the experimental data. This feature can be evaluated only a posteriori using the statistical information on the goodness of fit commonly provided by the software employed.

Two main problems may arise: The model is not able to predict the data, and the precisions of the parameter estimates are poor. The disagreement between data and model can be generally observed e.g. through a graphical comparison of the model curves with the experimental data. The graphical comparison may be subjective as it might be difficult to properly evaluate the influence of the experimental uncertainties on the goodness of fit; therefore, the evaluation of the ability of the model to predict the data should rely only on the available statistical indicators. Should these indicators suggest that the fit is unacceptable, it means that the model structure, although a priori identifiable, might be logically or physiologically wrong, and it needs to be changed. The precision of the parameter estimates depends primarily on the experimental uncertainties on the measured data and on the sampling schedule. As pointed out in Sect. 13.2.6, the choice of the time schedule for the collection of the biokinetic data is crucial to be able to properly characterise processes with different characteristic time constants. The uncertainties related to the parameter estimates can thus be improved, improving the precision of the measurements and/or optimizing the sampling schedule.

To summarise, an a priori identifiable model might be unable to describe the experimental data since its structure has been ill defined; alternatively, the model parameters may be uniquely but poorly defined. On the other hand, an excellent fit of the experimental data is not sufficient to provide the correct values of the model parameters if the model is not uniquely identifiable. The ideal combination of model and experimental setup is therefore the one that verifies both conditions of a priori identifiability (unique values for the unknown parameters) and a posteriori identifiability (the model solution closely resembles the characteristics of the data, and the parameters can be estimated with acceptable statistical precision).

13.4 Some Examples: Biokinetic Models Available in the Literature

Biokinetic models can look in many different ways depending on the aim of the model, the organs of interest and the data available. When choosing the number of compartments, there is no right or wrong; it all depends on the user and the aim of the study. As the statistician George Box said: "All models are wrong, but some are useful." The following are some examples of biokinetic models presented in the literature.

13.4.1 Two-Compartment Model for ^{131}I-G250 Antibody

Loh and coworkers [13] studied the biokinetics of ^{131}I-G250 antibody in renal cell carcinoma patients. They chose to use a two-compartment model representing the serum and the rest of the body (Fig. 13.7).

The activity was injected in the serum and assumed to be excreted from the rest of the body. During the time when the patients were isolated, due to radiation safety constraints, the whole-body activity content was obtained by survey

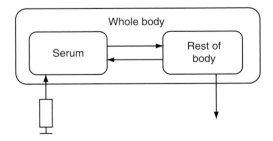

Fig. 13.7 The two-compartment model used by Loh and coworkers [13] when studying the biokinetics of ^{131}I-G250 antibody in patients with renal cell carcinoma. The activity was injected in the serum and excreted from the rest of the body

meter measurements using a handheld ionisation chamber. When the patients were released from isolation, planar whole body gamma camera images were collected. The geometric mean was taken of the anterior and posterior counts, and the patient thickness for attenuation correction was obtained from a CT image. Blood samples were drawn throughout the treatment, and the activity content was measured using a gamma counter.

13.4.2   Three- and Five-Compartment Models for ^{111}In-Labelled Monoclonal Antibody

Odom-Maryon and coworkers [14] studied the biokinetics of the ^{111}In-labelled monoclonal antibody T84.66 in patients with carcinoembryonic antigen-producing metastases. They used a three-compartmental model to describe the activity content in blood and the excretion with urine (Fig. 13.8, left panel) and a five-compartment model to describe the activity content in blood and urine as well as the liver, the rest of the body, and faeces (Fig. 13.8, right panel).

The activity was injected in the blood, and eight blood samples were taken: immediately after injection, 30 min after injection, and 1, 2 and 6 h after injections as well as at every acquisition. Four whole-body images were collected (6, 24, and 48 h after injection as well as one between 4 and 7 days postinjection). In addition, two SPECT images were collected: 48 h after injection and between 4 and 7 days postinjection. Daily urine samples were collected for up to 5 days after injection. A bowel cathartic was administered to the patients before every scan to reduce the faecal gastrointestinal activity, but the faeces were not collected. Blood and urine samples were measured in a gamma counter. The geometric mean of the anterior and posterior images was used to determine the activity in the organs and whole body.

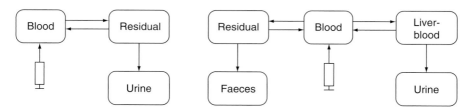

Fig. 13.8   The compartment models used by Odom-Maryon and coworkers [14] when studying the biokinetics of the ^{111}In-labelled monoclonal antibody T84.66 in patients with carcinoembryonic antigen-producing metastases. *Left:* The three-compartment model. *Right:* The five-compartment model

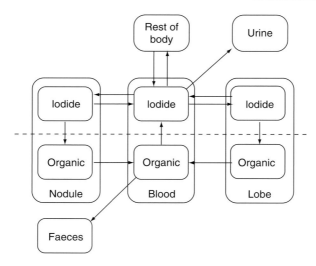

Fig. 13.9 The compartment model used by Janzen and coworkers [15] to describe the iodide uptake in the thyroid in patients with autonomous nodule syndrome

13.4.3 Nine-Compartment Model for Radioiodine

Janzen and coworkers [15] studied the biokinetics of radioiodine in the thyroid in patients with autonomous nodule syndrome; ^{123}I for dosimetric studies and ^{131}I for therapy was administered. The thyroids of the patients were imaged up to 120 h during the dosimetric study and up to 46 days during therapy. In some patients, also concentrations in blood and urine samples were measured. A nine-compartment model structure was developed (Fig. 13.9); the thyroid was split into a healthy lobe and an autonomously functioning nodule, each with its own characteristic kinetics. The structure explicitly differentiated between inorganic iodide and organic iodine in blood, nodule, and lobe, so that each of these tissues was actually described by two compartments. Organic iodide is excreted into the faeces, whereas inorganic iodine can be transported to other body tissues or excreted into the urine. A population kinetics approach was used to identify the parameters depending on individual kinetics and those common to the whole population and thus to single out those features of the biokinetic structure that need to be closely and individually investigated in each patient.

13.4.4 Thirteen-Compartment Model for ^{111}In-DOTATOC

Cremonesi and coworkers [16] used a compartment model containing 13 different compartments to describe the biokinetics of ^{111}In-DOTATOC in patients with neuroendocrine tumours. The compartments represented blood, extracellular fluid,

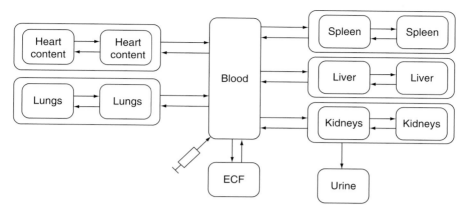

Fig. 13.10 The compartment model used by Cremonesi and coworkers [16] to describe the biokinetics of ^{111}In-DOTATOC in patients with neuroendocrine tumours. The activity was injected in the blood and excreted in the urine via the kidneys. The organs were represented by one fast and one slow component. *ECF* extracellular fluid

heart content, lungs, spleen, liver, kidneys, and urine (Fig. 13.10). The heart content, lungs, spleen, liver and kidneys were all represented by two compartments each. The two compartments for one organ correspond to one fast and one slow component.

The activity was injected in the blood, and blood samples were drawn at 30 s; 1, 3, 5, 10, 15, 20, and 30 min; and 1, 3, 12, 16, 24, 36, and 48 h after injection. All urine was collected up to 50 h after injection in eight different samples. The activity content in the blood and urine was determined with a gamma counter. Planar whole-body images were collected 30 min after injection as well as after 3–4, 24, and 48 h. One SPECT acquisition was made for each patient 3–4 h after injection. The conjugate view method was used to determine the activity content in the organs and whole body. In this study ^{111}In-DOTATOC was used to determine the biokinetics of ^{90}Y-DOTATOC due to the fact that ^{90}Y is not suitable for imaging with a gamma camera. This means that ^{111}In was used as a tracer for ^{90}Y, which was the tracee.

13.4.5 Twenty-Compartment Model for FDG

Hays and Segall [17] used a 20-compartment model to describe the biokinetics of the PET substance ^{18}F-FDG (2-[^{18}F]-2-deoxy-D-glucose). They used four compartments each for the myocardium, liver, and brain; three compartments for the lungs; two each for the whole body and other tissues; and one for the urine (Fig. 13.11).

The activity was injected in the blood, and dynamic PET images using a single bed position over the lower chest area, including the heart and the upper part of the liver, were collected. The data collection started immediately after injection. PET images were collected for 20 s each during the first 5 min, after that for 1 min each during the following 10 min, and for 5 min each for the following 75 min. Blood

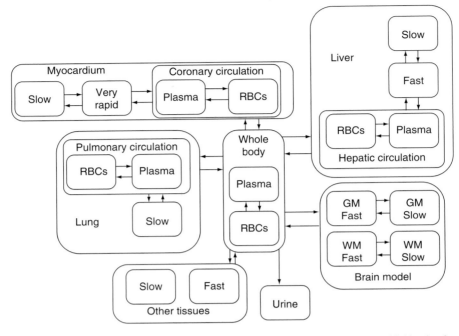

Fig. 13.11 The compartment model used by Hays and Segall [17] to describe the biokinetics for ^{18}F-FDG (2-[^{18}F]-2-deoxy-D-glucose). The activity was injected in the blood and excreted with the urine

samples were collected as close to the middle of each PET acquisition as possible. The imaging and blood sampling stopped after 90 min, and one urine sample was collected. Previously published parameters of the brain kinetics were used in the modelling.

References

1. Loevinger, R., Budinger, T., and Watson, E. *MIRD Primer for absorbed dose calculations.* Society of Nuclear Medicine, New York (1988).
2. ICRP. *Radiation dose to patients from radiopharmaceuticals.* ICRP Publication 53. Ann ICRP 18(1–4) (1987).
3. ICRP. *Radiation dose to patients from radiopharmaceuticals. Addendum 2 to ICRP Publication 53.* ICRP Publication 80. Ann ICRP 28(3) (1998).
4. ICRP. *Radiation dose to patients from radiopharmaceuticals. Addendum 3 to ICRP Publication 53.* ICRP Publication 106. Ann ICRP 38(1–2) (2008).
5. Bolch, W.E., Eckerman, K.F., Sgouros, G., and Thomas, S.R. *MIRD Pamphlet No. 21: a generalized schema for radiopharmaceutical dosimetry – standardization of nomenclature.* J. Nucl. Med. 50:477–484 (2009).
6. ICRP. *Nuclear decay data for dosimetric calculations.* ICRP Publication 107. Ann ICRP 38(3) (2008).

7. Siegel, J.A., Thomas, S.R., Stubbs, J.B., Stabin, M.G., Hays, M.T., Koral, K.F., Robertson, J.S., Howell, R.W., Wessels, B.W., Fisher, D.R., et al., *MIRD pamphlet no. 16: Techniques for quantitative radiopharmaceutical biodistribution data acquisition and analysis for use in human radiation dose estimates*. J Nucl Med. 40:37S–61S (1999).

8. Carson, E.R., Cobelli, C., and Finkelstein, L. *The mathematical modelling of metabolic and endocrine systems*. Wiley, New York (1983).

9. ICRP. *Limits for intakes of radionuclides by workers*. ICRP Publication 30 Part I. Ann ICRP 2(3–4) (1979), pp. 30–34.

10. ICRP. *Human alimentary tract model for radiological protection*. ICRP Publication 100. Ann ICRP 36(1–2) (2006).

11. Audoly, S., D'Angiò, L., Saccomani, M.P., and Cobelli, C. *Global identifiability of linear compartment models. A computer algebra algorithm*. IEEE Trans. Biomed. Eng. 45:36–47 (1998).

12. Bellù, G., Saccomani, M.P., Audoly, S., D'Angiò, L., *DAISY: a new software tool to test global identifiability of biological and physiological systems*. Comput. Meth. Progr. Biol. 88:52–61 (2007).

13. Loh, A., Sgouros, G., O'Donoghue, J.A., Deland, D., Puri, D., Capitelli, P., Humm, J.L., Larson, S.M., Old, L.J. and Divgi, C.R. *Pharmacokinetic model of iodine-131-G250 antibody in renal cell carcinoma patients*. J. Nucl. Med. 39:484–489 (1998).

14. Odom-Maryon, T.L., Williams, L.E., Chai, A., Lopatin, G., Liu, A., Wong, Y.C., Chou, J., Clarke, K.G., and Raubitschek, A.A. *Pharmacokinetic modeling and absorbed dose estimation for chimeric anti-CEA antibody in humans*. J. Nucl. Med. 38:1959–1966 (1997).

15. Janzen, T., Giussani, A., Canzi, C., Gerundini, P., Oeh, U., and Hoeschen, C. Investigation of biokinetics of radioiodine with a population kinetics approach. Radiat Prot. Dosim. 139:232–235 (2010).

16. Cremonesi, M., Ferrari, M., Zoboli, S., Chinol, M., Stabin, M.G., Orsi, F., Maecke, H.R., Jermann, E., Robertson, C., Fiorenza, M., et al. *Biokinetics and dosimetry in patients administered with (111)In-DOTA-Tyr(3)-octreotide: implications for internal radiotherapy with (90)Y-DOTATOC*. Eur. J. Nucl. Med. 26:877–886 (1999).

17. Hays, M.T. and Segall, G.M., *A mathematical model for the distribution of fluorodeoxyglucose in humans*. J. Nucl. Med. 40:1358–1366 (1999).

Chapter 14
Voxel Phantoms for Internal Dosimetry

Maria Zankl, Helmut Schlattl, Nina Petoussi-Henss, and Christoph Hoeschen

14.1 Introduction

The calculation of radiation dose from internally incorporated radionuclides is based on so-called absorbed fractions (AFs) and specific absorbed fractions (SAFs). AFs specify the fraction of energy emitted by radioactivity in a given (source) organ that is absorbed in the source organ itself and in other (target) organs. SAFs are AFs divided by target organ mass. According to the MIRD (Medical Internal Radiation Dose) formalism [1], the equations relating absorbed dose, S values, and SAFs are as follows:

$$D(r_T, T_D) = \sum_{r_S} \tilde{A}(r_S, T_D) \cdot S(r_T \leftarrow r_S) \tag{14.1}$$

where $D(r_T, T_D)$ is the absorbed dose in target region r_T, $\tilde{A}(r_S, T_D)$ is the time-integrated or cumulated activity in source region r_S (equal to the total number of transformations in the source region occurring during time T_D), and $S(r_T \leftarrow r_S)$ is the so-called S value (i.e., absorbed dose in the target region per unit of cumulated activity in the source region in $mGyMBq^{-1}s^{-1}$). This equation takes into account the biokinetic processes; that is, it considers the fact that a target region is generally irradiated by several source regions. For each source and target region combination,

$$S(r_T \leftarrow r_S) = k \cdot \sum_{i} E_i \cdot Y_i \cdot \Phi(r_T \leftarrow r_S, E_i) \tag{14.2}$$

where E_i is the mean energy of radiation type i (i.e., the mean energy of the ith transition), and Y_i is the yield of radiation type i per transformation (i.e., the number of ith transitions per nuclear transformation); $\Phi(r_T \leftarrow r_S, E_i)$ (acronym SAF) is the fraction of energy emitted in r_S that is absorbed in r_T divided by the

M. Zankl (✉), H. Schlattl, N. Petoussi-Henss, and C. Hoeschen
Department of Medical Radiation Physics and Diagnostics, Helmholtz Zentrum
München – German Research Center for Environmental Health, Ingolstädter Landstr.1,
85764 Neuherberg, Germany
e-mail: zankl@helmholtz-muenchen.de

M.C. Cantone and C. Hoeschen (eds.), *Radiation Physics for Nuclear Medicine*,
DOI 10.1007/978-3-642-11327-7_14, © Springer-Verlag Berlin Heidelberg 2011

corresponding target region mass; k is a proportionality constant for the conversion between the different units used. This sum takes into account the physical properties of the radionuclide decay (i.e., the different radiation types and energies emitted in the nuclear transformations and their respective emission probabilities). Note: In the equations of this chapter, the nomenclature recently proposed in MIRD Pamphlet 21 [2] is adopted.

Equations 14.1 and 14.2 show that for the assessment of internal patient doses, a variety of parameters has to be known: the amount and path of intake of the radionuclide, the temporal distribution of the radionuclide within the body, the nuclear decay data of the radionuclide considered, and the energy-dependent SAFs for the relevant source and target regions. The methods for assessing the first-mentioned parameters are covered by other chapters of this book. The present chapter concentrates on the evaluation of SAFs and presents some examples of absorbed doses per unit activity administered for selected radiopharmaceuticals. Finally, possibilities regarding a personalised (patient-specific) dosimetry are discussed.

14.2 The Calculation of Specific Absorbed Fractions

14.2.1 Phantoms

For over 30 years and until recently, SAFs were calculated at the Oak Ridge National Laboratory (ORNL) on the basis of MIRD-type anthropomorphic phantoms, which are mathematical phantoms describing the geometry of a reference body and its organs by mathematical inequalities [3–5]. A whole range of SAF data exists, covering all ages [6]. More recently, three-dimensional (3D) models of the human body were constructed from computed tomographic (CT) or magnetic resonance imaging (MRI) data of real persons. Among other laboratories, the Helmholtz Zentrum München – German Research Center for Environmental Health (i.e., the former GSF – National Research Center for Environment and Health) has developed 12 voxel phantoms of individuals of different stature and ages: 2 paediatric ones and 4 male and 6 female adult models [7–11]. An overview of the currently existing voxel models can be found in the literature [12, 13]. This type of body model, the so-called voxel model (from voxel = volume element), was found to represent human anatomy more realistically than the MIRD-type (also called "mathematical") models with respect to organ shape and location. First studies revealed that, due to the simplified inequalities used to describe the organs in the MIRD-type models, some of the interorgan distances tended to be larger in these models than they are in reality, where neighbouring organs are often in immediate contact. This leads, especially for lower photon energies, to higher values of SAFs for many source–target organ combinations for the voxel models. These differences range from tens of percent to orders of magnitude [14–22]. For internal dosimetry, the influencing parameters are the relative position of source and target organs (for organ cross fire) and organ mass (for organ self-absorption) [14, 16, 20, 23]. Consequently, voxel phantoms that are based on CT data of real persons could significantly contribute to improved dose assessment for patients.

Hence, the International Commission on Radiological Protection (ICRP) and the International Commission of Radiation Units and Measurements (ICRU) decided to use voxel models for the update of organ dose conversion coefficients, following the recent revision of the ICRP recommendations [24]. According to the ICRP, these voxel models should be representative of the adult reference male and reference female [25] with respect to their external dimensions and their organ masses. Several attempts have been made to create voxel-based phantoms that correspond to the ICRP reference anatomical values [9, 14, 26–31]. The reference computational phantoms adopted finally by the ICRP and ICRU were developed at the Helmholtz Zentrum München in collaboration with the ICRP Task Group Dose Calculations (DOCAL). Approximately 140 organs and tissues were segmented, including objects that were not previously contained in the MIRD-type phantoms, such as the main blood vessels, cartilage, muscles, and lymphatic nodes. The external dimensions and nearly all organ masses of both models were adjusted to the ICRP reference values. A detailed description of the reference computational phantoms was provided by the ICRP [32]. The resulting reference computational phantoms are characterised briefly in the following discussion.

The adult reference computational phantoms are defined in the form of a 3D voxel array. The array entries are organ identification numbers that describe to which organ a specific voxel belongs. The columns correspond to the x coordinates, with numbers increasing from right to left; the rows correspond to the y coordinates, increasing from front to back; and the slices correspond to the z coordinates, increasing from the toes to the vertex of the body. The main characteristics of the adult reference computational models are summarised in Table 14.1.

A graphical representation of the male and female adult reference computational phantoms is shown in Fig. 14.1.

The skeleton is a highly complex structure composed of cortical bone, trabecular bone, red and yellow bone marrow, cartilage, teeth, and "miscellaneous" skeletal tissues (i.e., blood vessels and periosteum). A subregion of the bone marrow located within 50 μm from the bone surfaces is defined as *endosteal tissues* (previously also referred to as "bone surfaces"). The internal dimensions of most of these tissues are smaller than the resolution of the reference computational phantoms; thus, these volumes could not be segmented in the reference computational phantoms.

Table 14.1 Main characteristics of the adult male and female reference computational phantoms

Property	Male	Female
Mass (kg)	73	60
Height (m)	1.76	1.63
Voxel in-plane resolution (mm)	2.137	1.775
Slice thickness (mm)	8	4.84
Approximate number of tissue voxels (millions)	1.9	3.9
Number of columns	254	299
Number of rows	127	137
Number of slices	220	346

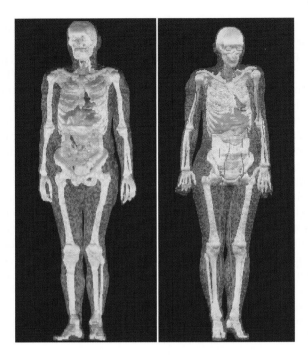

Fig. 14.1 Frontal view of the male (*left*) and female (*right*) reference computational phantoms representing the ICRP (International Commission on Radiological Protection) adult reference male and reference female

Nevertheless, an attempt was made to represent the gross spatial distribution of the source and target volumes in the voxel models as realistically as possible at the given voxel resolution. Therefore, the skeleton was divided into those 19 bones and bone groups for which individual data on red bone marrow content and marrow cellularity were given in ICRP Publication 70 [33]. These were subdivided as cortical bone, spongiosa, and – if applicable – medullary cavity. In total, 48 different identification numbers were assigned to the skeleton. To fully exploit the available data on red bone marrow distribution and marrow cellularity [33], the classified spongiosa regions of different bones are composed of different ratios of mineral bone and red and yellow bone marrow [10, 32].

It is not possible to segment the entire blood pool of the body; the larger part of the blood volume is situated in the small vessels and capillaries inside the organs and cannot be segmented. On the other hand, the elemental compositions of tissues as listed in the relevant ICRP and ICRU publications [25, 34] are exclusive of blood. Therefore, the blood content in each organ as given in ICRP Publication 89 [25] was considered by including a blood portion in the elemental tissue composition of each organ.

In addition to the skeletal fine structures, there are other regions that could not be directly segmented or could not be adjusted to their reference values due to their small size or complex structure. These are as follows:

- Extrathoracic airways (represented by an entire voxel layer lining the airways of nose, larynx, and pharynx, whereas in reality they are only a few tens of micrometres thick)
- Bronchi (not followed further than the first generations of branching)
- Bronchioles (too small for segmentation; represented by homogeneous lung tissue with a density that is the average of the densities of bronchiolar tissue and air)
- Skin (represented by one voxel layer wrapping the exterior of the phantoms)
- Cartilage (only partly segmented; rest included in elemental composition of spongiosa regions)
- Gallbladder (sum of wall and contents has reference mass sum; otherwise, not enough wall voxels to encompass the entire volume of the content)
- Adipose tissue ("residual tissue" used to adjust whole body mass to reference value; matching of reference adipose tissue mass approximately achieved)

The eye lens masses were adjusted to the reference values, but the shape of the lenses is somewhat coarse due to their small size compared to the voxel dimensions.

14.2.2   Monte Carlo Calculations

SAFs for a variety of phantoms have been calculated at various sites. For the new male and female reference computational phantoms, SAFs were calculated at the Helmholtz Zentrum München using the well-known Monte Carlo radiation transport program package EGSnrc [35]. The calculations were performed for 55 source and more than 65 target regions. For photons and electrons, 25 particle energies between 10 and 10 MeV were considered, and the number of histories followed per simulation was 5–10 million.

14.2.3   Comparison of SAF Values Calculated with Stylized/Reference Computational Phantoms (Photons)

In the following, SAF values calculated using the male and female reference computational phantoms are compared with SAF values from the ORNL [6, 36, 37].

For organ self-absorption, there is a strong dependence of SAFs on organ mass, which results in rather small differences between the MIRD values and the voxel SAF values. Examples are shown in Fig. 14.2, which presents SAF values for self-absorption in the liver and the thyroid.

There is excellent agreement between the values for the reference computational phantoms and the values calculated previously at the ORNL. For the thyroid, a rather small organ, the values calculated with EGSnrc and full secondary particle transport are slightly lower for photon energies above approximately 500 keV than the

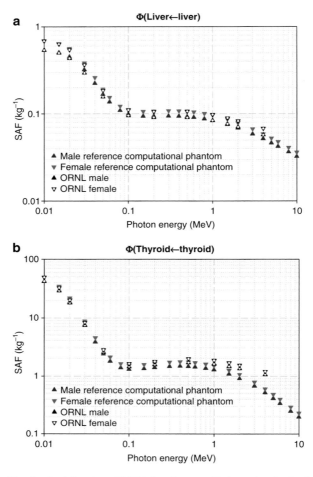

Fig. 14.2 Specific absorbed fractions (SAFs) for photon self-absorption in the liver (a) and the thyroid. (b). *ORNL*: Oak Ridge National Laboratory

previous values calculated with kerma approximation. This shows that, for small organs, secondary particle equilibrium is not established above this photon energy.

For organ cross fire, larger discrepancies occur for some organ pairs between the reference voxel and MIRD-type models, as can be seen in Fig. 14.3; these are due to the fact that interorgan distances tend to be larger in the MIRD-type phantoms than in reality. Consequently, the SAF values then have a tendency to be lower for the MIRD-type phantoms than for voxel phantoms.

Figure 14.4 presents photon SAF values for irradiation of the stomach wall by a uniformly distributed source in the stomach contents. Here again, excellent agreement of the EGSnrc calculated values for the new ICRP/ICRU reference computational phantoms with the previous values from the ORNL can be seen.

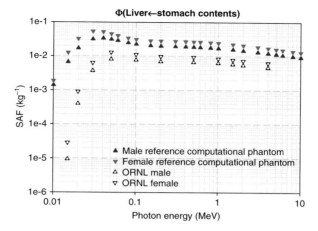

Fig. 14.3 Specific absorbed fractions (SAFs) for photon cross fire from stomach contents as source region to the target region liver. *ORNL* Oak Ridge National Laboratory

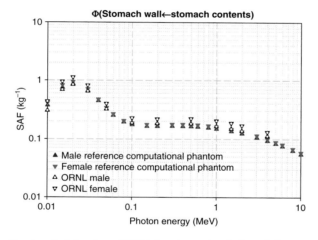

Fig. 14.4 Specific absorbed fractions (SAFs) for photon irradiation of the stomach wall from stomach contents as source region. *ORNL* Oak Ridge National Laboratory

14.2.4 Comparison of ICRP 30 Approximations and Calculated Electron SAF Values

As mentioned, the photon SAF values for the mathematical models, used as input data from many software tools for internal dosimetry, were calculated at the ORNL [6]. For electrons, no SAF values stemming from particle transport were available. Instead, the following assumptions were used [38]: For solid regions,

$$\phi(r_T \leftarrow r_S, E_i) = \begin{cases} 1, & \text{if } r_T = r_S \\ 0, & \text{if } r_T \neq r_S \\ \frac{M_{r_T}}{M_{\text{WB}}}, & \text{if } r_S = \text{WB} \end{cases} \quad (14.3)$$

where r_T is the target region; r_S is the source region; WB is the whole body; M_{r_T} and M_{WB} are the masses of the target region and of the whole body, respectively; and $\phi(r_T \leftarrow r_S, E_i)$ is the fraction of energy emitted in r_S that is absorbed in r_T (AF).

For contents of walled organs, the assumption was used that the dose to the wall is the dose at the surface of a half-space or half the equilibrium dose to the contents:

$$\Phi(\text{wall} \leftarrow \text{contents}) = \frac{0.5}{M_{\text{contents}}} \quad (14.4)$$

which means that the AF is half the mass ratio of wall and contents,

$$\phi(\text{wall} \leftarrow \text{contents}) = 0.5 \cdot \frac{M_{\text{wall}}}{M_{\text{contents}}} \quad (14.5)$$

More recently, electron AFs have been published by the ICRP for irradiation of the radiation-sensitive target cell layers in specific regions of the alimentary tract from sources in the wall and contents of these regions [39]. Therefore, the approximation mentioned of half the equilibrium dose to the contents is currently retained only for the urinary bladder wall.

In the following, SAF values for electrons calculated using the male and female reference computational phantoms are compared with SAF values following the assumptions of ICRP Publication 30 [38]. These results confirm earlier findings that the previously applied assumption of electrons being fully absorbed in the source organ itself is not always true at electron energies above approximately 300–500 keV [23].

Electron SAF values for self-absorption in the liver and the thyroid are shown in Fig. 14.5. It can be seen that high-energy electrons have the ability to leave the source organ. This effect is much more pronounced in the thyroid, which is a small organ and therefore has a high surface-to-volume ratio.

Monte Carlo calculated electron SAFs for cross fire from the kidneys to the thyroid and from the stomach contents to the liver are shown in Fig. 14.6. While the electron cross-fire SAFs for distant organs, such as kidneys and thyroid, are indeed negligibly small, the electron SAFs for neighbouring organs, such as stomach contents and liver, can reach the same magnitude as those for photons for electron energies above 1 MeV.

Finally, electron SAFs calculated with EGSnrc for irradiation of the urinary bladder wall by activity in the contents are presented in Fig. 14.7. The energy-independent values from the ICRP 30 assumption are 2.5 kg^{-1} for both phantoms if a urinary bladder content mass of 0.2 kg is assumed. Obviously, this is a substantial overestimation, especially for electron energies below 1 MeV.

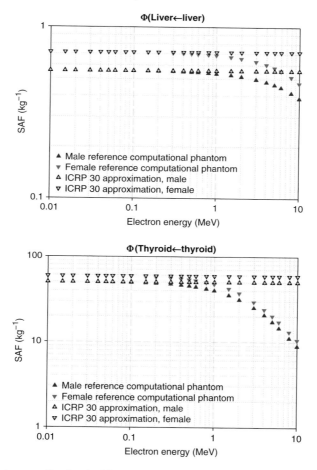

Fig. 14.5 Electron specific absorbed fractions (SAFs) calculated with the adult male and female reference computational phantoms for self-absorption in the liver and the thyroid, compared with the constant values derived using the ICRP 30 approximation of full electron absorption in the source organ

14.3 Evaluation of Absorbed Doses per Administered Activity for Selected Radiopharmaceuticals

For the following absorbed dose coefficients, photon as well as electron SAFs were calculated and implemented for the calculation of doses due to radiopharmaceuticals. As an example, absorbed doses per administered activity of ^{18}F-FDG (2-fluoro-2-deoxy-d-glucose) are given in Table 14.2, evaluated with and without explicit electron transport, to quantify the impact of the improved electron dosimetry. For comparison, also the values from ICRP Publication 106 are given that have been evaluated with MIRD-type phantoms and without electron transport.

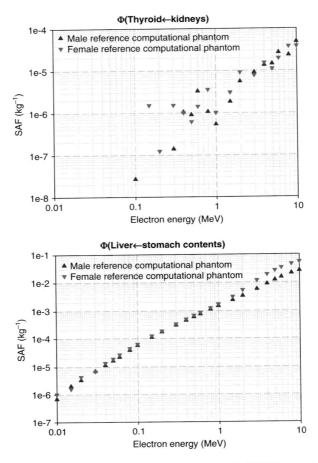

Fig. 14.6 Electron specific absorbed fractions (SAFs) calculated with EGSnrc and the adult male and female reference computational phantoms for cross fire from the kidneys to the thyroid and from the stomach contents to the liver. For both cases, the International Commission on Radiological Protection (ICRP) 30 assumption of full electron absorption in the source organ results in zero SAFs

The organ-absorbed doses per administered activity for the two evaluation methods agreed within 4% for all non-source organs. As expected, the organ-absorbed doses for most of the source organs evaluated with calculated electron SAFs were slightly lower than those for the ICRP 30 approach, with differences up to 4.1% for the solid source organs. The dose to the urinary bladder wall was 61% lower for the male and 59% lower for the female reference computational phantom using the calculated electron SAFs, compared to the ICRP 30 approach, due to the large difference in SAFs, especially at electron energies below 1 MeV (see Fig. 14.7). The fact that most of the differences are so small is probably because of the predominance of activity cumulated in the remaining tissues. Therefore, the decrease in source organ

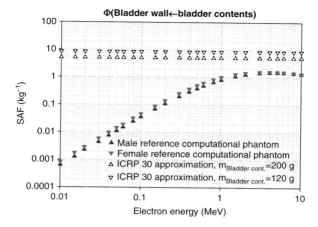

Fig. 14.7 Electron specific absorbed fractions (SAFs) calculated with EGSnrc and the adult male and female reference computational phantoms for cross fire from the urinary bladder contents to the urinary bladder wall. The ICRP 30 assumption of half the equilibrium dose to the contents results in an energy-independent value of $2.5\,\mathrm{kg}^{-1}$ if one assumes a urinary bladder contents mass of $0.2\,\mathrm{kg}$, which is the value for both reference computational phantoms. If a urinary bladder contents mass of $0.120\,\mathrm{kg}$ is assumed, the value of the SEECAL software [6], the overestimation is even more pronounced. *ICRP* International Commission on Radiological Protection

self-dose and increase in organ cross-fire dose for neighbouring organs have only a small impact on the overall absorbed dose values.

14.4   Towards a More Personalised Dosimetry

While both the MIRD-type and the new ICRP/ICRU reference computational phantoms were designed to match the physical characteristics of the ICRP reference male and reference female [25], they do not necessarily represent the individual patient under study. Voxel phantoms, based on CT data of real persons, could significantly contribute to a better approach in providing guidance on how to properly scale reference phantom S values to fit the individual patient. A study using seven individual adult voxel phantoms was therefore performed to examine the influence of individual organ masses and individual organ topology on SAF and S values [20, 41].

14.4.1   Characteristics of Seven Adult Voxel Phantoms

For this study, a variety of voxel models was used, all except one segmented at the Helmholtz Zentrum München. The main characteristics of these phantoms are given in Table 14.3 and selected organ masses in Table 14.4.

Table 14.2 Comparison of organ-absorbed doses per administered activity (mGy/MBq) for the ^{18}F-FDG (2-fluoro-2-deoxy-d-glucose) radiopharmaceutical, calculated using the ICRP 30 approximation and electron specific absorbed fractions (SAFs). The MIRD values of ICRP Publication 53 are also shown [40]. The source organs are marked by an asterisk

	Organ	Organ dose per unit administered activity (mGy MBq^{-1})				
		Male		Female		ICRP
		ICRP 30 approximation	With electron SAFs	ICRP 30 approximation	With electron SAFs	Publication 53
	R-marrow	1.25E-2	1.25E-2	1.44E-2	1.44E-2	1.1E-2
	Colon	1.27E-2	1.26E-2	1.59E-2	1.58E-2	1.4E-2
	Lungs	1.28E-2	1.28E-2	1.50E-2	1.51E-2	1.1E-2
	St-wall	1.18E-2	1.17E-2	1.28E-2	1.27E-2	1.2E-2
	Breast	8.71E-3	8.70E-3	1.14E-2	1.13E-2	1.1E-2
	Ovaries			2.41E-2	2.40E-2	1.5E-2
	Testes	1.06E-2	1.06E-2			1.5E-2
*	UB-wall	1.66E-1	6.45E-2	1.67E-1	6.82E-2	1.7E-1
	Oesophagus	1.42E-2	1.43E-2	1.62E-2	1.65E-2	
	Liver	1.17E-2	1.17E-2	1.33E-2	1.33E-2	1.2E-2
	Thyroid	9.82E-3	9.85E-3	1.18E-2	1.18E-2	9.7E-3
	Endost-BS	1.02E-2	1.03E-2	1.22E-2	1.22E-2	1.0E-2
*	Brain	2.55E-2	2.54E-2	2.80E-2	2.79E-2	2.6E-2
	S-glands	8.89E-3	8.90E-3	1.12E-2	1.12E-2	
	Skin	7.67E-3	7.44E-3	9.21E-3	8.89E-3	
	Adrenals	1.27E-2	1.28E-2	1.42E-2	1.42E-2	1.4E-2
	ET	9.83E-3	9.86E-3	1.18E-2	1.18E-2	
	GB-wall	1.10E-2	1.08E-2	1.28E-2	1.25E-2	
*	Ht-wall	5.74E-2	5.54E-2	7.34E-2	7.05E-2	6.5E-2
*	Kidneys	2.06E-2	2.05E-2	2.30E-2	2.29E-2	2.1E-2
	Lymph	1.39E-2	1.40E-2	1.51E-2	1.53E-2	
	Muscle	9.87E-3	9.87E-3	1.20E-2	1.20E-2	
	O-mucosa	9.48E-3	9.52E-3	1.11E-2	1.12E-2	
	Pancreas	1.18E-2	1.18E-2	1.33E-2	1.33E-2	1.2E-2
	Prostate	2.97E-2	3.01E-2			
	SI-wall	1.41E-2	1.40E-2	1.75E-2	1.75E-2	1.3E-2
	Spleen	1.16E-2	1.16E-2	1.32E-2	1.32E-2	1.2E-2
	Thymus	1.19E-2	1.19E-2	1.47E-2	1.48E-2	
	Uterus			3.46E-2	3.48E-2	2.0E-2

R-marrow: red marrow, St-wall: stomach wall, UB-wall: urinary bladder wall, Endost-BS: endosteum, S-glands: salivary glands, ET: extrathoracic airways, GB-wall: gallbladder wall, Ht-wall: heart wall, O-mucosa: oral mucosa, SI-wall: small intestine wall

14.4.2 Method to Adjust Source Organ Mass to Reference Values

Since photons have only a moderate ability to penetrate, especially at the low energies typical for many radionuclides, a large amount of energy is absorbed in the source organ itself. Hence, the SAF values for self-absorption are strongly influenced by the source organ mass, which can be quite different for the individual

Table 14.3 Main characteristics of seven individual voxel phantoms. Voxelman is the model from Zubal et al. [42]. All other voxel models were segmented at the Helmholtz Zentrum München

	Age (years)	Height (cm)	Mass (kg)	Body region	Voxel width (mm)	Voxel depth (mm)	Slice thickness (mm)	No. of slices
Frank	48			Head and trunk	0.74	0.74	5	193
Golem	38	176	69	Whole body	2.08	2.08	8	220
Visible Human	38	180	103	Head to knee	0.91	0.94	5	250
Voxelman		178	70	Head to thigh	3.75	3.75	4	236
Donna	40	176	79	Whole body	1.875	1.875	10	179
Helga	26	170	81	Head to thigh	0.98	0.98	10	114
Irene	32	163	51	Whole body	1.875	1.875	5	348

phantoms. For studying the influence of the individual organ topology, independent of the individual organ masses, this influence of the source organ mass is impedimental. Therefore, a method was developed that allows performing the calculations as if the source organ had the ICRP reference mass, although in reality this is not the case. Thus, two independent calculations could be performed for each phantom and source organ, one using the individual source organ mass, the other assuming the reference organ mass [20].

14.4.3 Dependence of SAFs for Photons on Individual Anatomical Characteristics

The organ with the largest mass variability among the seven voxel models was the thyroid, ranging from 6.2 g (Voxelman) to 31.8 g (Visible Human), that is, approximately a factor of 5. The large influence of the organ mass can be seen in Fig. 14.8, which shows photon SAFs for self-absorption in the thyroid for individual as well as reference thyroid masses.

The MIRD Committee recommends scaling reference SAF values for organ self-absorption with the inverse 2/3 power of the organ mass [43]. By examining the ratio of the individual thyroid self-absorption SAFs calculated with individual thyroid masses and those assuming reference thyroid masses, this relation was confirmed [41] (see Fig. 14.9).

For organ cross fire, on the other hand, no dependence on source organ mass was found. This can be seen in Fig. 14.10, in which SAFs for irradiation of the lungs by activity in the thyroid for the seven phantoms with their original, individual thyroid mass are shown, as well as the ratios of the SAFs evaluated with the individual thyroid masses over those evaluated with the reference thyroid masses. While the SAFs for the individual phantoms showed a large variability, the SAF ratios were close to unity for all phantoms and photon energies above 100 keV. This shows that the SAFs for organ cross-fire do not depend on the source organ mass.

Table 14.4 Selected organ masses of the seven individual voxel phantoms compared with the reference values [25]

	ICRP reference male	Frank	Golem	Visible Human	Voxelman	ICRP reference female	Donna	Helga	Irene
Adipose tissue	13,500	30,590[a]	19,970	26,040[a]	26.7	17,700	34,820	39,800[a]	11,630
Adrenals	14	13.6	22.8	7.2	3.7	14.	21.7	6.6	12.4
Bladder wall	45	56.6	68.4	51.9	186	45	61.0	60.8	39.0
Bladder contents	200	218	272	41.2	382	200	45.0	22.1	25.5
Brain	1,400	1,827	1,218	1,429	1,081	1,200	1,208	1,279	1,255
Breast	26	1.8[b]	–	–	–	360	43.9[b]	134[b]	57.0[b]
Colon wall	370	379	297	790	1,080	360	322	426	271
Colon contents	355	666	237	2,186	67.0	294	309	609	273
Gallbladder	10	9.0	8.3	3.1	19.4	8	6.6	5.7	19.3
Heart	330[c]	381[c]	716	637	553	240[c]	446[c]	531[c]	472[c]
Kidneys	310	494	316	383	450	275	281	390	212
Liver	1,800	2,072	1,592	2,037	1,729	1,400	1,585	1,757	1,225
Lungs	1,000	1,338	729	1,026	912	800	631	463	685
Muscle	28,000	13,820[a]	26,970	40,970[a]	21,160[a]	17,000	25,420	21,340[a]	21,100
Oesophagus	40	62.6	30.1	86.2	37.9	30	27.7	28.0	24.3
Ovaries	–	–	–	–	–	11	12.1	11.9	11.9
Pancreas	100	60.0	71.9	62.5	46.8	85	41.2	43.3	61.9
Red bone marrow	1,170	1,363[a]	1,177	1,399	1,223	900	1,012	1,043	916
Skeleton	10,500	7,250[a]	10,450	8,841[a]	6,448[a]	7,800	7,484	6,503[a]	8,201
Skin	2,600	737[a]	4,703	1,950[a]	18,000	1,790	4,351	1,653[a]	3,620
Small intestine	640	664	959[d]	521	1,562[d]	600	435	443	396
+cont.	400	382		767		331	363	637	311
Spleen	180	339	174	266	329	150	306	298	203
Stomach wall	150	127	233	258	303	140	195	62.8	163
Stomach contents	250	177	140	166		207	305	10.3	205
Testes	35	–	21.1	25.5	102	–	–	–	–
Thymus	20	3.3	10.7	14.0	–	20	19.0	7.7	25.3
Thyroid	20	22.3	25.8	31.8	6.2	17	18.7	31.5	20.0
Uterus	–	–	–	–	–	80	71.7	79.8	25.0

ICRP International Commission on Radiological Protection.

[a] Corresponding mass of part of the arms and legs is missing.

[b] Glandular tissue only.

[c] Wall (muscle) only.

[d] Wall and contents not separated.

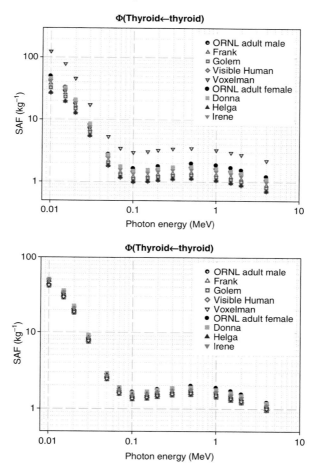

Fig. 14.8 Photon specific absorbed fraction (SAF) values for self-absorption in the thyroid for seven individual voxel phantoms. *Top*: Individual thyroid masses ranging from 6.2 to 31.8 g. *Bottom*: Reference thyroid masses of 17 g (female phantoms) and 20 g (male phantoms), respectively. *ORNL* Oak Ridge National Laboratory

For cross-fire photon SAFs, it is further known that there is reciprocity between source and target organs [3, 44], which means that

$$\Phi(r_T \leftarrow r_S, E_i, t) \approx \Phi(r_S \leftarrow r_T, E_i, t) \tag{14.6}$$

This ensures that the cross-fire photon SAFs are also independent of the target organ mass. Hence, the reason for the individual variability must be due to the topology of the organs, namely, the distances between the source and target organ.

Table 14.5 shows interorgan distances for selected organ pairs in the seven adult voxel phantoms compared to those for the MIRD-type phantom Adam [45].

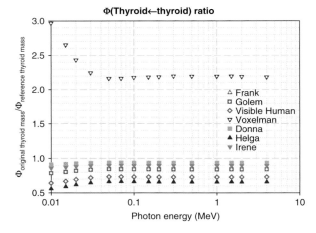

Fig. 14.9 Ratios for thyroid self-absorption of specific absorbed fraction (SAF) values for seven adult voxel models calculated using individual thyroid masses, over SAFs calculated for thyroid masses adjusted to the reference value. The values of the inverse 2/3 power of the organ mass ratios for the various phantoms are 0.93 (Frank), 0.84 (Golem), 0.73 (Visible Human), 2.18 (Voxelman), 0.93 (Donna), 0.66 (Helga), and 0.90 (Irene)

Although there is a large variability of the interorgan distances among the seven individual voxel phantoms, it can be seen that, in many cases, the interorgan distance for the MIRD-type phantom was larger than any of the interorgan distances for the voxel phantoms.

14.4.4 Organ Absorbed Doses per Incorporated Activity for Individual Voxel Phantoms

In the following, organ-absorbed doses per incorporated activity for the seven individual voxel models are given in Table 14.6 for ^{99m}Tc-pertechnetate and in Table 14.7 for ^{123}I-iodide, two examples of frequently used radiopharmaceuticals. Accepting the low impact of Monte Carlo calculated electron SAFs demonstrated, the ICRP 30 approach was used for this evaluation.

Compared to photon SAFs, which may differ by orders of magnitude, the variability of organ-absorbed doses is much reduced. Nevertheless, there were significant dose differences among the individual voxel models: In only 27 of 44 cases, the doses for the voxel models agreed within 50% of their average value; in 10 cases, the variation exceeded 60%, and 100% in 4 cases. Despite the relatively large dose ranges spanned by the voxel models, in 19 cases the doses for the MIRD phantom were still outside this range. Since for the evaluation of organ-absorbed doses all source organs were assigned reference masses, these results showed that there is a significant influence of individual organ topology on the resulting doses. An additional evaluation of the organ-absorbed doses per incorporated activity for

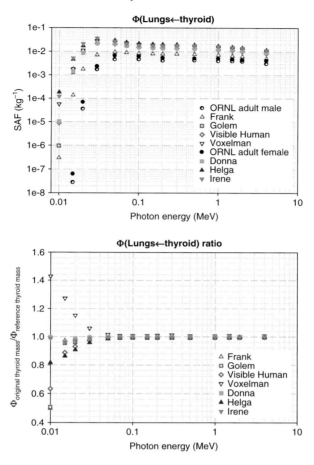

Fig. 14.10 Specific absorbed fractions (SAFs) for irradiation of the lungs by activity in the thyroid, calculated with the original, individual thyroid mass (*top*) and ratios of these SAFs over those evaluated with thyroid masses adjusted to the reference values (*bottom*). *ORNL* Oak Ridge National Laboratory

[123]I using the original, individual source organ masses increased the variability of absorbed doses in these organs significantly, with the largest increase from a (variation of 17.2% to 203.8%) for the thyroid [20].

14.5   Summary/Conclusions

The ICRP and the ICRU have decided to use voxel-based models for the update of organ dose conversion coefficients, following the recent revision of the ICRP recommendations. Adult male and female reference computational phantoms were

Table 14.5 Interorgan distances (cm) for the voxel models of this study and the adult male Medical Internal Radiation Dose (MIRD) phantom. Each distance shown is the arithmetic mean of the distances of 5 million point pairs randomly selected in the two organs considered

Organ 1	Organ 2	MIRD	Frank	Golem	Visible human	Voxelman	Donna	Helga	Irene
Bladder contents	Liver	31.2	29.3	32.1	36.2	28.8	32.8	35.2	28.6
Bladder contents	Lungs	46.2	39.2	43.2	47.5	39.6	43.2	44.7	43.1
Bladder contents	Stomach	28.7	27.9	31.5	36.7	29.1	31.9	35.7	30.3
Bladder contents	Thyroid	64.3	54.2	54.7	59.2	50.7	53.9	54.2	53.9
Kidneys	Bladder wall	27.7	20.2	25.7	27.6	26.4	27.8	28.4	25.6
Kidneys	Liver	15.3	15.5	13.4	15.5	14.1	12.5	13.9	11.1
Kidneys	Lungs	24.2	23.4	20.9	23.6	20.6	18.8	19.6	20.2
Kidneys	Stomach	15.4	15.5	13.5	15.6	13.6	11.8	12.9	11.1
Kidneys	Thyroid	41.8	37.1	31.3	34.5	30.5	28.6	28.2	30.6
Liver	Lungs	21.3	18.4	16.5	17.2	17.5	15.7	15.4	17.9
Liver	Stomach	18.8	14.2	13.3	15.1	14.1	13.8	14.5	12.2
Liver	Thyroid	36.9	29.4	25.3	25.9	25.8	23.7	22.1	27.3
Stomach contents	Lungs	22.3	18.4	16.7	16.1	15.9	16.8	14.5	15.7
Stomach contents	Thyroid	38.0	28.5	25.3	24.5	24.1	24.4	20.3	24.2
Thyroid	Lungs	22.6	19.3	15.2	15.9	14.9	14.9	13.5	14.5

developed at the Helmholtz Zentrum München in collaboration with the ICRP Task Group DOCAL. Approximately 140 organs and tissues were segmented, including objects that were not previously contained in the MIRD-type phantoms. The external dimensions and nearly all organ masses of both models were adjusted to the ICRP reference values.

Photon SAFs have been calculated for the new voxel-based reference computational phantoms adopted by the ICRP and the ICRU. For photon SAFs for organ self-absorption, there is excellent agreement between the values for the reference computational phantoms and the values calculated previously at the ORNL. For organ cross fire, larger discrepancies occur for some organ pairs between the reference voxel and MIRD-type models; these are due to the fact that interorgan distances tend to be larger in the MIRD-type phantoms than in reality. Consequently, the SAF values then have a tendency to be lower for the MIRD-type phantoms than for voxel phantoms. For irradiation of the wall by activity in the contents of an organ, again excellent agreement of the EGSnrc calculated values for the new ICRP/ICRU reference computational phantoms with the previous values from the ORNL can be seen.

The explicit Monte Carlo calculation of electron SAF values showed that high-energy electrons have the ability to leave the source organ. Consequently, the ICRP 30 approach assuming full absorption of electrons in the source organ presents an overestimation for organ self-absorption and an underestimation for organ cross fire for electron energies above approximately 1 MeV. For neighbouring organs, such as stomach contents and liver, electron SAFs can reach the same magnitude as those for photons for electron energies above 1 MeV. For irradiation of the urinary bladder

Table 14.6 Organ-absorbed doses per incorporated activity of ^{99m}Tc-pertechnetate for seven individual voxel phantoms and the corresponding values for Medical Internal Radiation Dose (MIRD)-type phantoms from ICRP Publication 53 [40]. The source organs are marked by an asterisk. In case of the bladder, the source is in the contents

Organ	Organ dose per unit administered activity (mGy MBq^{-1})							
	MIRD	Frank	Golem	Visible Human	Voxelman	Donna	Helga	Irene
Bladder wall *	1.9E-02	1.54E-02	1.31E-02	1.96E-02	2.20E-02	1.97E-02	2.10E-02	2.36E-02
Breast	2.3E-03	–	–	–	–	2.42E-03	2.09E-03	2.96E-03
Colon wall *	4.2E-02	4.38E-02	4.10E-02	4.10E-02	3.89E-02	5.01E-02	4.66E-02	5.07E-02
Ovaries	1.0E-02	–	–	–	–	7.70E-03	6.11E-03	9.60E-03
Testes	2.7E-03	–	2.53E-03	2.45E-03	4.44E-03	–	–	–
Liver	3.9E-03	6.08E-03	5.86E-03	5.16E-03	5.22E-03	4.61E-03	6.15E-03	5.34E-03
Lungs	2.7E-03	4.84E-03	3.84E-03	3.66E-03	4.35E-03	3.26E-03	3.30E-03	3.83E-03
Oesophagus	–	6.04E-03	4.53E-03	5.47E-03	5.37E-03	3.88E-03	4.11E-03	3.77E-03
Red bone marrow	6.1E-03	3.54E-03	3.80E-03	3.34E-03	5.96E-03	4.04E-03	3.98E-03	4.64E-03
Skeleton	3.9E-03	4.87E-03	3.95E-03	4.54E-03	6.39E-03	4.63E-03	5.44E-03	5.01E-03
Skin	–	1.50E-03	1.89E-03	1.78E-03	8.22E-03	2.73E-03	2.04E-03	2.98E-03
Stomach wall *	2.9E-02	2.85E-02	3.00E-02	2.94E-02	3.69E-02	3.02E-02	3.44E-02	3.08E-02
Thyroid *	2.3E-02	2.35E-02	2.88E-02	2.39E-02	2.87E-02	3.38E-02	2.79E-02	2.73E-02
Adrenals	3.6E-03	6.79E-03	5.75E-03	5.06E-03	5.18E-03	8.82E-03	7.71E-03	6.83E-03
Brain	–	2.33E-03	1.85E-03	1.70E-03	1.93E-03	1.83E-03	1.93E-03	2.45E-03
Kidneys *	5.0E-03	8.73E-03	7.14E-03	6.94E-03	7.77E-03	6.78E-03	8.73E-03	6.76E-03
Muscle	–	2.89E-03	2.93E-03	3.29E-03	3.60E-03	4.00E-03	4.14E-03	4.21E-03
Pancreas	5.9E-03	8.02E-03	1.04E-02	7.52E-03	8.16E-03	9.73E-03	1.02E-02	9.15E-03
Small intestine wall *	1.8E-02	1.31E-02	1.13E-02	9.88E-03	1.38E-02	1.35E-02	1.27E-02	1.38E-02
Spleen	4.4E-03	6.52E-03	5.42E-03	5.35E-03	7.42E-03	7.91E-03	7.93E-03	6.80E-03
Thymus	–	1.97E-03	3.48E-03	2.69E-03	–	3.17E-03	2.53E-03	4.04E-03
Uterus	8.1E-03	–	–	–	–	9.98E-03	6.45E-03	1.33E-02

wall by activity in the contents, the energy-independent values from the ICRP 30 assumption present a substantial overestimation, especially for electron energies below 1 MeV.

The organ-absorbed doses per administered activity evaluated with and without explicit electron transport for ^{18}F-FDG agree within a few percent for all non-source organs. As expected, the organ-absorbed doses for the source organs evaluated with calculated electron SAFs are slightly lower than those for the ICRP 30 approach. The dose to the urinary bladder wall is significantly lower for the male and female reference computational phantom using the calculated electron SAFs, compared to the ICRP 30 approach, due to the large difference in SAFs, especially at electron energies below 1 MeV. However, the decrease in source organ self-dose and increase in organ cross-fire dose for neighbouring organs have only a small impact on the overall absorbed dose values, apart from the urinary bladder wall.

Table 14.7 Organ-absorbed doses per incorporated activity of ^{123}I-iodide for seven individual voxel phantoms and the corresponding values for Medical Internal Radiation Dose (MIRD)-type phantoms from ICRP Publication 53 [40]. The source organs are marked by an asterisk. In case of the bladder, the source is in the contents

Organ	Organ dose per unit administered activity (mGy MBq^{-1})							
	MIRD	Frank	Golem	Visible Human	Voxelman	Donna	Helga	Irene
Bladder wall *	6.90E-02	5.38E-02	4.83E-02	7.73E-02	4.62E-02	7.67E-02	8.73E-02	9.83E-02
Breast	5.00E-03	–	–	–	–	5.35E-03	4.24E-03	6.90E-03
Colon wall *	1.50E-02	1.26E-02	1.31E-02	1.36E-02	2.00E-02	1.30E-02	1.24E-02	1.34E-02
Ovaries	1.10E-02	–	–	–	–	1.30E-02	1.44E-02	2.12E-02
Testes	5.20E-03	–	4.90E-03	4.74E-03	8.20E-03	–	–	–
Liver	6.30E-03	1.07E-02	9.73E-03	8.42E-03	1.04E-02	9.00E-03	1.01E-02	9.51E-03
Lungs	6.10E-03	1.15E-02	1.21E-02	1.17E-02	1.41E-02	1.09E-02	1.27E-02	1.39E-02
Oesophagus		3.12E-02	5.44E-02	3.41E-02	4.50E-02	4.29E-02	5.65E-02	5.09E-02
Red bone marrow	9.80E-03	7.90E-03	8.65E-03	7.18E-03	1.01E-02	8.66E-03	9.40E-03	1.06E-02
Skeleton	7.50E-03	1.16E-02	9.25E-03	9.97E-03	1.38E-02	1.03E-02	1.33E-02	1.19E-02
Skin	4.10E-03	4.24E-03	4.16E-03	3.74E-03	1.52E-02	6.06E-03	4.55E-03	6.58E-03
Stomach wall *	6.80E-02	1.65E-01	1.73E-01	1.77E-01	2.09E-01	1.76E-01	1.97E-01	1.79E-01
Thyroid *	3.2	3.08	3.08	3.09	3.1	3.65	3.59	3.59
Adrenals	6.40E-03	1.30E-02	1.15E-02	8.34E-03	1.41E-02	1.45E-02	2.14E-02	1.57E-02
Brain	5.70E-03	4.72E-03	3.99E-03	3.67E-03	4.43E-03	4.21E-03	4.47E-03	5.17E-03
Kidneys *	1.10E-02	1.65E-02	1.37E-02	1.27E-02	1.90E-02	1.37E-02	1.56E-02	1.36E-02
Muscle	7.90E-03	7.55E-03	6.94E-03	7.25E-03	8.32E-03	8.44E-03	9.58E-03	8.91E-03
Pancreas	1.40E-02	1.75E-02	1.95E-02	1.44E-02	2.61E-02	2.10E-02	2.50E-02	2.53E-02
Small intestine wall *	4.30E-02	4.97E-02	3.67E-02	3.98E-02	5.57E-02	5.07E-02	4.72E-02	5.21E-02
Spleen	9.60E-03	1.26E-02	1.25E-02	1.16E-02	1.83E-02	1.70E-02	1.65E-02	1.66E-02
Thymus	7.10E-03	4.84E-03	4.56E-02	9.68E-02	–	1.49E-02	1.77E-02	1.16E-02
Uterus	1.40E-02	–	–	–	–	2.45E-02	1.63E-02	4.65E-02

While both the MIRD-type and the new ICRP/ICRU reference computational phantoms were designed to match the physical characteristics of the ICRP reference male and reference female, they do not necessarily represent the individual patient under investigation. A study using seven individual adult voxel phantoms was performed to examine the influence of individual organ masses and individual organ topology on SAF and S values.

The MIRD Committee recommends scaling reference SAF values for organ self-absorption with the inverse 2/3 power of the organ mass; this relation was confirmed. For organ cross fire, on the other hand, no dependence on source organ mass was found. Together with the reciprocity principle, this ensures that the cross-fire photon SAFs are also independent of the target organ mass. Hence, the reason for the individual variability must be due to the variability of the interorgan distances.

Compared to photon SAFs, which may differ by orders of magnitude, the variability of organ-absorbed doses per administered activity of radiopharmaceuticals is much reduced. Nevertheless, there are significant dose differences among the individual voxel models. Despite the relatively large dose ranges spanned by the voxel models, in many cases the doses for the MIRD phantom are still outside this range. This shows that there is a significant influence of individual organ topology on the resulting doses. An additional evaluation of the organ-absorbed doses per incorporated activity using the original, individual source organ masses showed a significant mass dependence for those organs where self-absorption is dominant.

It should, however, be noted that the cumulated activity in the source organs is ruled by biokinetic processes and body metabolism. Here, the uncertainties are much higher than for the calculated SAF values, and the individual variability may be much higher. Therefore, steps towards individual dosimetry are only meaningful if individual biokinetic data also are available.

References

1. Loevinger, R., Budinger, T. and Watson, E. MIRD primer for absorbed dose calculations. (1988).
2. Bolch, W. E., Eckerman, K. F., Sgouros, G. and Thomas, S. R. MIRD Pamphlet No. 21: A generalized schema for radiopharmaceutical dosimetry – standardization of nomenclature. J. Nucl. Med. 50, 477–484 (2009).
3. Cristy, M. and Eckerman, K. F. Specific absorbed fractions of energy at various ages from internal photon sources, Part I: Methods. TM-8381/V1. (1987).
4. Snyder, W. S., Ford, M. R. and Warner, G. G. Estimates of specific absorbed fractions for monoenergetic photon sources uniformly distributed in various organs of a heterogeneous phantom. MIRD Pamphlet No. 5, Revised. (1978).
5. Snyder, W. S., Ford, M. R., Warner, G. G. and Fisher, H. L. Estimates of absorbed fractions for monoenergetic photon sources uniformly distributed in various organs of a heterogeneous phantom. Medical Internal Radiation Dose Committee (MIRD). Pamphlet No. 5. J. Nucl. Med. 10(Suppl. 3), 7–52 (1969).
6. Cristy, M. and Eckerman, K. F. SEECAL 2.0. Program to calculate age-dependent specific effective energies. ORNL/TM-12351 (1993).
7. Fill, U., Zankl, M., Petoussi-Henss, N., Siebert, M. and Regulla, D. Adult female voxel models of different stature and photon conversion coefficients for radiation protection. Health Phys. 86, 253–272 (2004).
8. Petoussi-Henss, N., Zankl, M., Fill, U. and Regulla, D. The GSF family of voxel phantoms. Phys. Med. Biol. 47, 89–106 (2002).
9. Zankl, M., Becker, J., Fill, U., Petoussi-Henß, N. and Eckerman, K. F. GSF male and female adult voxel models representing ICRP Reference Man – the present status. Proceedings of the Monte Carlo Method: Versatility Unbounded in a Dynamic Computing World. Chattanooga, TN (2005).
10. Zankl, M., Eckerman, K. F. and Bolch, W. E. Voxel-based models representing the male and female ICRP reference adult - the skeleton. Radiat. Prot. Dosim. 127, 174–186 (2007).
11. Zankl, M. and Wittmann, A. The adult male voxel model "Golem" segmented from whole body CT patient data. Radiat. Environ. Biophys. 40, 153–162 (2001).
12. Zaidi, H. and Xu, X. G. Computational anthropomorphic models of the human anatomy: The path to realistic Monte Carlo modeling in radiological sciences. Ann. Rev. Biomed. Eng. 9, 471–500 (2007).

13. Xu, X. G. and Eckerman, K. F. Handbook of Anatomical Models for Radiation Dosimetry. (Boca Raton, London, New York: Taylor & Francis) (2010) ISBN 978 1 4200 5979 3.

14. Jones, D. G. A realistic anthropomorphic phantom for calculating specific absorbed fractions of energy deposited from internal gamma emitters. Radiat. Prot. Dosim. 79, 411–414 (1998).

15. Petoussi-Henss, N. and Zankl, M. Voxel anthropomorphic models as a tool for internal dosimetry. Radiat. Prot. Dosim. 79, 415–418 (1998).

16. Smith, T., Petoussi-Henss, N. and Zankl, M. Comparison of internal radiation doses estimated by MIRD and voxel techniques for a 'family' of phantoms. Eur. J. Nucl. Med. 27, 1387–1398 (2000).

17. Yoriyaz, H., Santos, A. D., Stabin, M. G. and Cabezas, R. Absorbed fractions in a voxel-based phantom calculated with the MCNP-4B code. Med. Phys. 27, 1555–1562 (2000).

18. Smith, T. J., Phipps, A. W., Petoussi-Henss, N. and Zankl, M. Impact on internal doses of photon SAFs derived with the GSF adult male voxel phantom. Health Phys. 80, 477–485 (2001).

19. Stabin, M. G. and Yoriyaz, H. Photon specific absorbed fractions calculated in the trunk of an adult male voxel-based phantom. Health Phys. 82, 21–44 (2002).

20. Zankl, M., Petoussi-Henss, N., Fill, U. and Regulla, D. The application of voxel phantoms to the internal dosimetry of radionuclides. Radiat. Prot. Dosim. 105, 539–548 (2003).

21. Petoussi-Henss, N., Zankl, M. and Nosske, D. Estimation of patient dose from radiopharmaceuticals using voxel models. Cancer Biother. Radiopharm. 20, 103–109 (2005).

22. Petoussi-Henss, N., Li, W. B., Zankl, M. and Eckerman, K. F. SEECAL utilizing voxel-based SAFs. Radiat. Prot. Dosim. 127, 214–219 (2007).

23. Chao, T. C. and Xu, X. G. Specific absorbed fractions from the image-based VIP-Man body model and EGS4-VLSI Monte Carlo code: internal electron emitters. Phys. Med. Biol. 46, 901–927 (2001).

24. ICRP. The 2007 Recommendations of the International Commission on Radiological Protection. ICRP Publication 103 (2007).

25. ICRP. Basic anatomical and physiological data for use in radiological protection: reference values. ICRP Publication 89. Ann ICRP 32(3–4) (2002).

26. Dimbylow, P. J. The development of realistic voxel phantoms for electromagnetic field dosimetry. Proceedings of Workshop on Voxel Phantom Development. Chilton, UK, pp. 1–7 (1996).

27. Jones, D. G. A realistic anthropomorphic phantom for calculating organ doses arising from external photon irradiation. Radiat. Prot. Dosim. 72, 21–29 (1997).

28. Kramer, R., Vieira, J. W., Khoury, H. J., Lima, F. R. A. and Fuelle, D. All about MAX: A male adult voxel phantom for Monte Carlo calculations in radiation protection dosimetry. Phys. Med. Biol. 48, 1239–1262 (2003).

29. Kramer, R., Khoury, H. J., Vieira, J. W., Loureiro, E. C. M., Lima, V. J. M., Lima, F. R. A. and Hoff, G. All about FAX: a female adult voxel phantom for Monte Carlo calculations in radiation protection dosimetry. Phys. Med. Biol. 49, 5203–5216 (2004).

30. Dimbylow, P. Development of the female voxel phantom, NAOMI, and its application to calculations of induced current densities and electric fields from applied low frequency magnetic and electric fields. Phys. Med. Biol. 50, 1047–1070 (2005).

31. Kramer, R., Khoury, H. J., Vieira, J. W. and Lima, V. J. M. MAX06 and FAX06: update of two adult human phantoms for radiation protection dosimetry. Phys. Med. Biol. 51, 3331–3346 (2006).

32. ICRP. Adult reference computational phantoms. ICRP Publication 110 (2009).

33. ICRP. Basic anatomical and physiological data for use in radiological protection: the skeleton. ICRP Publication 70. Ann ICRP 25(2) (1995).

34. ICRU. Photon, electron, proton and neutron interaction data for body tissues. ICRU Report 46 (1992).

35. Kawrakow, I. and Rogers, D. W. O. The EGSnrc code system: Monte Carlo simulation of electron and photon transport. PIRS Report No. 701 (2003).

36. Cristy, M. and Eckerman, K. F. Specific absorbed fractions of energy at various ages from internal photon sources, Part V: Fifteen-year-old male and adult female. TM-8381/V5 (1987).

37. Cristy, M. and Eckerman, K. F. Specific absorbed fractions of energy at various ages from internal photon sources, Part VII: Adult male. TM-8381/V7. (1987).
38. ICRP. Limits for intakes of radionuclides by workers. Part 1. ICRP Publication 30 (1979).
39. ICRP. Human alimentary tract model for radiological protection. ICRP Publication 100. Ann ICRP 36(1–2) (2006).
40. ICRP Radiation dose to patients from radiopharmaceuticals. ICRP Publication 53. Ann ICRP 18(1–4) (1987).
41. Petoussi-Henss, N., Bolch, W. E., Zankl, M., Sgouros, G. and Wessels, B. Patient-specific scaling of reference S-values for cross-organ radionuclide S-values: what is appropriate? Radiat. Prot. Dosim. 127, 192–196 (2007).
42. Zubal, I. G., Harrell, C. R., Smith, E. O., Rattner, Z., Gindi, G. and Hoffer, P. B. Computerized three-dimensional segmented human anatomy. Med. Phys. 21, 299–302 (1994).
43. Snyder, W. S., Ford, M. R., Warner, G. G. and Watson, E. E. "S" absorbed dose per unit cumulated activity for selected radionuclides and organs. MIRD Pamphlet 11, Revised (1975).
44. Cristy, M. Applying the reciprocal dose principle to heterogeneous phantoms: Practical experience from Monte Carlo studies. Phys. Med. Biol. 28, 1289–1303 (1983).
45. Kramer, R., Zankl, M., Williams, G. and Drexler, G. The calculation of dose from external photon exposures using reference human phantoms and Monte Carlo methods, Part I: The male (Adam) and female (Eva) adult mathematical phantoms. GSF Report S-885 (1982).

Index